Optimisation of Corrosion Control for Lead in Drinking Water Using Computational Modelling Techniques

Metals and Related Substances in Drinking Water: Research Report Series

Optimisation of Corrosion Control for Lead in Drinking Water Using Computational Modelling Techniques

Authors

Dr. C. R. Hayes and Dr. T. N. Croft

Collaborators

A. Campbell, City of Ottawa Water (CA)

I. P. Douglas, City of Ottawa Water (CA)

P. Gadoury, Providence Water (US)

M. R. Schock, US Environmental Protection Agency (US)

IWA Publishing

London • New York

Published by

IWA Publishing
Alliance House
12 Caxton Street
London SW1H 0QS, UK
Telephone: +44 (0)20 7654 5500
Fax: +44 (0)20 7654 5555
Email: publications@iwap.co.uk
Web: www.iwapublishing.com

First published 2013

British Library Cataloguing in Publication Data
A CIP catalogue record for this book is available from the British Library

ISBN: 9781780404783 (Paperback)
ISBN: 9781780404790 (eBook)

Contents

Foreword

In many respects, lead in drinking water has become a forgotten problem, since the mid 1980s when a range of environmental controls were implemented to reduce exposure to lead. This is largely because the sampling protocols, that underpin regulatory controls, are mostly inadequate and have tended to under-estimate the amount of lead that can be present in drinking water (IWA, 2010). Meanwhile, the World Health Organization has published a booklet on Childhood Lead Poisoning (WHO, 2010) that concludes that there is no safe level of lead exposure for children and that all environmental sources could and should be eliminated.

In December 2006, an international research network on "metals and related substances in drinking water" was established with funding from COST (www.cost.eu), an institution within the European Union. Over its four year life-span it built up representation from 27 European countries and established links with Canada, the US, the European Commission's Joint Research Centre and the World Health Organization. It was successful in convening four international conferences in which lead in drinking water was a major topic. These conferences enabled the results of numerous research and practitioner experiences to be assimilated, creating a much better understanding of the extent and nature of the problems from lead in drinking water in both Europe and North America, culminating in the publication of a Best Practice Guide (IWA, 2010).

In November 2010, this international research network became a Specialist Group within the International Water Association. It continues to be active on the topic of lead in drinking water and is committed to stimulating collaborative research and the dissemination of knowledge. In mid 2011, the scope for a possible collaborative project was identified concerning the optimisation of corrosion control for lead in drinking water using computational modelling techniques. A "proof-of-concept" project commenced in December 2011 and was concluded in April 2012. The aim was to demonstrate that optimisation techniques developed in the United Kingdom could be applied with equal success to water supply systems in Canada and the United States. The findings of the project are presented in this report.

This report is the first in the Research Report Series that is being published by the IWA Specialist Group on Metals and Related Substances in Drinking Water.

Acknowledgements

The Authors would like to express their thanks to the collaborators for their input to this project, without which the case studies would not have been possible. The project has succeeded in bringing together researchers and practitioners from across the Atlantic with the common aim of continuing to tackle the lead in drinking water problem. All parties contributed their time and resources without charge, thereby avoiding the constraints of external funding and allowing the project to develop in a flexible way as it progressed.

Thanks are due to Clayton Commons, from Rhode Island Health, for the provision of information on sequential sampling surveys.

Thanks are also due to the Review Panel:

Matthew Bower, Operations Team, Drinking Water Quality Regulator for Scotland (UK)
Professor Michael Ritchie Moore, Water Quality Research Australia (AU)
Dr Simoni Triantafyllidou, Virginia Tech (US)

Disclaimers

Whilst every reasonable attempt has been made to present the information in this report in a fair and balanced manner, the reader should none-the-less satisfy themselves of its relevance to their specific circumstances. It must also be appreciated that some aspects of the topic of plumbosolvency control do not enjoy total consensus of opinion and that practices have varied around the world.

The mention of specific companies or of certain manufacturers' products does not imply that they are endorsed or recommended by the publisher or authors in preference to others of a similar nature that are not mentioned. All reasonable precautions have been taken by the authors to verify the information contained in this publication. However, the published material is being distributed without warranty of any kind, either express or implied. The responsibility for the interpretation and use of the material lies with the reader. In no event shall the publisher or authors be liable for damages arising from its use. The views expressed by the authors do not necessarily represent the decisions or the stated policies of any organization referred to in this publication.

The U.S. Environmental Protection Agency, through its Office of Research and Development, funded and managed, or partially funded and collaborated in, the research described herein. It has been subjected to the Agency's administrative review and has been approved for external publication. Any opinions expressed in this report are those of the authors and do not necessarily reflect the views of the Agency, therefore, no official endorsement should be inferred. Any mention of trade names or commercial products does not constitute endorsement or recommendation for use.

Executive Summary

Computational modelling

The research employed two types of computational modelling. The first type was a compliance model that was used to investigate the optimisation of plumbosolvency control on the basis of the sampling protocols used in Canada and the US. It used a single pipe model within a probabilistic framework to predict lead emissions across a water supply system and a sampling model was then used to predict compliance with the relevant standards for drinking water. The single pipe model was also used to investigate sequential sampling at an individual home. The second type was a computational fluid dynamics (CFD) model that was used to investigate laminar flow along a pipe.

Objective

Compliance modelling has been used to very good effect in the optimisation of plumbosolvency control in the United Kingdom, particularly in the optimisation of orthophosphate dosing. Over 100 water supply systems have been modelled, involving 30% of the UK's water companies. This "proof-of-concept" project had the overall objective of demonstrating that these modelling techniques could also be applicable to the circumstances of Canada and the United States, via three case studies.

Overall conclusions

Modelling was shown to have an important potential role in the optimisation of plumbosolvency control, in the context of both Canada and the United States, by being able to link plumbosolvency control treatment conditions to system-wide compliance, on the basis of the 30 minutes and 6+ hours stagnation sampling protocols that variously apply. Modelling was also shown to be able to support investigations at individual homes and demonstrated potential limitations in the use of sequential sampling as a survey method.

Calibrating the compliance model

In compliance modelling for a water supply system, the most important calibration data are: (i) the lengths and diameters of the lead service lines; (ii) the types (copper, brass, etc), lengths and diameters of premise plumbing; (iii) the plumbosolvency of the water; and (iv) the percentage of homes with a lead service line in the water supply system. The extent of this data varied in the three case studies necessitating the use of some assumptions, which were adjusted (to a limited extent) in order to match predicted compliance results to those observed. Obtaining adequate calibration data in the future on pipework characteristics should not be particularly onerous or costly, particularly if opportunities to undertake brief inspections of service connections and premise plumbing were taken during sampling campaigns. Plumbosolvency characteristics, which strongly influence the results of sampling for lead in drinking water, can be quickly determined by laboratory testing.

Premise plumbing

Premise plumbing characteristics were found to be critical to the results of stagnation sampling for lead. The water stood in non-lead pipework (most commonly made of copper or iron) adjacent to the faucet, is not representative of the water that has stood in a lead service line. As the volume of premise plumbing was found to normally exceed one litre, most first litre samples drawn after flushing and stagnation therefore under-estimate the extent of build up of lead in the domestic pipework system supplying drinking water to a home. This effect was clearly discernible in the results of both sequential sampling (as undertaken by utilities of other agencies) and modelling.

Sequential sampling

CFD analysis and compliance modelling both demonstrated the skewing and lowering of lead results from sequential sampling under laminar flow conditions, consistent with observations in the three case study cities. Modelling indicated that the profile of lead concentrations obtained from sequential sampling was strongly influenced by the length and diameter of both the lead pipe and non-lead pipe, as well as by the flow condition (plug or laminar). As these characteristics will vary from house to house, the use of sequential sampling will not provide a reliable survey basis for optimising plumbosolvency control or for assessing regulatory compliance, unless the sampling locations remain constant. Routine (e.g. monthly) sequential sampling at a small number of selected houses could provide a basis for assessing the performance of plumbosolvency control measures, if sampling flows were normalised.

Case study conclusions

Compliance modelling gave a preliminary indication that there was scope to further reduce lead concentrations in City A by orthophosphate dosing, although operational feasibility would first need to be confirmed. It also indicated that the current plumbosolvency control treatment measures in City B were close to optimum by pH elevation alone. A diagnostic assessment of water quality data, sequential survey results and the history of the installation of lead service lines, when taken together, suggested that lead in drinking water may be an issue in City C that warrants further prompt investigation, which could be supported to good effect by compliance modelling.

Regulatory aspects

There is scope to improve the various sampling protocols that are used in the regulation of lead in drinking water in both Canada and the United States. In relation to Canada, the research revealed inconsistencies between the use of the 30 minutes and 6+ hours stagnation sampling protocols. Compliance modelling also indicated that Tier 2 compliance was much more stringent than Tier 1, if benchmarked by the same standard for lead in drinking water. In relation to the US, the possible inclusion of sequential sampling in a revised Lead Copper Rule must be considered very carefully, recognising how results may be influenced by pipework characteristics. In both Canada and the US, the continued use of first draw one litre samples after stagnation for regulatory assessment must be questioned. A solution to these problems could be to adopt a more holistic risk based approach, in which sampling was only a part. Implicit in risk assessment would need to be an understanding of the interactions between water supply quality and pipework, aspects that can be greatly informed by compliance modelling.

Chapter 1

Introduction

1.1 CORRECTIVE WATER TREATMENT FOR REDUCING LEAD IN DRINKING WATER

The major focus of the research project was the optimisation of the corrective treatment measures that are most commonly used for reducing lead in drinking water. In low alkalinity water supplies, plumbosolvency can be greatly reduced by pH elevation up to 10.0 although the additional dosing of a corrosion inhibitor may be necessary in some cases. In high alkalinity waters, pH elevation may cause problems with calcite precipitation and so it has become common practice to rely on the dosing of a corrosion inhibitor.

The most effective corrosion inhibitor for reducing lead solubility is orthophosphate which converts some of the lead carbonate, within the internal corrosion deposits of lead pipes, to lead phosphate which has a lower solubility. This conversion depends on the alkalinity of the water and the dose of orthophosphate. Lead solubility can also be affected by natural organic matter, particularly the humic and fulvic acids associated with coloured upland waters. Further details on corrective water treatment measures can be found in IWA's Best Practice Guide for the Control of Lead in Drinking Water (IWA, 2010).

Regardless of the details of the corrosion chemistry involved, it is possible to describe the plumbosolvency characteristics of drinking water in numerical terms for modelling purposes. This involves defining the concentrations of lead that will dissolve from the internal surface of a lead pipe for any contact time period. The greater the plumbosolvency of the water, the higher will be the lead concentration for any particular contact time.

1.2 REGULATORY BACKGROUND IN THE UNITED STATES AND THE NEED FOR FURTHER OPTIMISATION OF PLUMBOSOLVENCY CONTROL

The Lead Copper Rule (LCR) was issued in 1991 as part of the implementation of the Safe Drinking Water Act 1986 (SDWA). It is implemented at State level, mainly through State regulations, as approved by the responsible US Environmental Protection Agency Regional Office for that jurisdiction.

The LCR sets a Maximum Contaminant Level Goal of zero for lead and an Action Level for lead of 15 μg/l. Compliance with the Action Level for lead is based on the 90th percentile concentration determined from sampling of homes identified to be at highest risk of elevated lead based on documented occurrence of lead service lines and leaded solder. A one litre, first draw sample is collected from the cold water (normally in the kitchen) tap considered to be the principal source of water for human consumption. It is collected from the selected homes after a minimum of 6 hours of standing (stagnation). Surveys must be undertaken twice every six months from designated water supply areas, the number of samples in a survey ranging from 5 to 100 for standard monitoring and 5 to 50 for reduced monitoring, for populations ranging from <100 to $>100{,}000$, respectively.

Tiers of risk level for high lead must be determined for homes in each water supply area to determine sampling sites. These are as follows:

- *Tier 1*: single family residential houses with a lead service line or with lead soldered plumbing installed after 1982;
- *Tier 2*: multiple family residential houses or other building types with the same plumbing characteristics as Tier 1, but only if an insufficient number of Tier 1 houses are available;
- *Tier 3*: single family residences with lead soldered copper plumbing installed before 1983, but only if an insufficient number of Tier 1 and 2 houses/buildings are available.

LCR compliance samples must be taken from homes of the lowest possible tier, that is, Tier 2 homes are sampled only if a sufficient number of Tier 1 homes are not available, and Tier 3 homes are sampled only if a sufficient number of Tier 1 and 2 homes are not available.

If a water supply area fails to comply with the Action Level for lead then optimal corrosion control treatment must be determined and then implemented, and a programme of lead service line replacement for pipe sections owned by the public water system must be subsequently initiated at a minimum rate of 7% per annum.

For large water systems, the LCR requires that the 90th percentile lead level must reflect "optimal corrosion control treatment," which is defined under the regulation as the lowest possible lead levels without violating other SDWA primary regulations. The frequency of exceeding the 15 μg/l 90th percentile Action Level will trigger other activities, such as public notification and partial lead service line replacement.

There are a range of issues relating to the effectiveness of the LCR for lead (IWA, 2010):

(1) Detailed interpretations vary between the States, particularly on what is considered to be optimum corrosion control;
(2) The criteria for selecting sampling locations enable houses without lead service lines (up to 50%) to be included in the surveys that are undertaken, making it easier for compliance with the Action Level for lead to be achieved (in this context, it is likely that the criteria based on lead solder are less relevant);
(3) The first draw one litre samples taken will comprise or include any water stood in non-lead pipe-work, in-situ between a lead service line and the tap outlet, effectively diluting any lead in the water that has stood in the lead service line, again making it easier for compliance with the Action Level for lead to be achieved (it is common practice for pipe-work to be fully flushed just prior to the stagnation period).
(4) Recurring access to the same houses for sampling after 6+ hours of stagnation is difficult to achieve both for each monitoring period, and for continuation of the monitoring site pool over time. This introduces variation into the sampling pool and therefore potential variation in survey results.
(5) The sampling protocol is not amenable to direct sampling by trained water utility staff, or their contractors and relies on the competence and diligence of the house-holders (not only their good will), which may vary.
(6) Some water utilities have compromised optimal plumbosolvency and lead release control because of difficulties in meeting other drinking water regulatory standards.

Presently, the LCR is undergoing review, and changes in requirements may emerge at the end of the process. Clearly, there is scope for improving the way in which lead in drinking water is regulated and it would be beneficial to achieve a better consensus on the definition of the term "optimisation" as it applies to plumbosolvency control. The common difficulty is the absence of a simple, rapid control loop that water utilities can use to relate treatment conditions to lead emissions at consumers' faucets, due principally to sampling limitations. These issues are the focus of this research project.

1.3 REGULATORY BACKGROUND IN CANADA AND THE NEED FOR FURTHER OPTIMISATION OF PLUMBOSOLVENCY CONTROL

The Guidelines for Canadian Drinking Water Quality (Health Canada, 1992), state a Maximum Allowable Concentration of 10 μg/l for lead, based on samples taken after thorough flushing. It remains the health-based lead limit for drinking water in Canada. Flushing prior to sampling will, in most cases, result in any lead that has built up within lead service lines being missed, making this standard an ineffective basis for health protection.

However, Health Canada (2009) has since developed technical guidance on controlling corrosion in distribution systems which complements the 1992 Guideline. It outlines the sampling methodologies and actions to be taken to reduce potential exposure to lead from drinking water resulting from corrosion. The guidance provides two options for sampling from residential properties.

Option 1:

The first option is intended for systems in which lead service lines (LSL) are present and where a minimum of 50% of the sampling sites target residences with a LSL. It is based on:

(1) A first sampling tier (Tier 1) relating to the 1st litre drawn from the faucet after 6+ hour's stagnation, similar to the US LCR; corrective measures should be implemented and additional sampling conducted if 15 μg/l is exceeded at more than 10% of the sites sampled.
(2) A second sampling tier (Tier 2) where utilities conduct additional sampling (sequential sampling of the 2nd, 3rd and 4th litres, taken after stagnation) at 10% of the sites sampled in Tier 1 with high lead concentrations (taken to mean above 15 μg/l), or at all sites where Tier 2 samples have been collected.

This Tier 2 sampling is intended to provide the lead emission profile of the home plumbing system and to determine the source(s) of lead, and help to better inform the approaches to be considered for corrosion control.

Option 2:

The second option is intended for systems in which only residences with lead service lines are sampled and where there are difficulties in 6+ hours stagnation sampling. It is based on taking 4 one litre samples sequentially after flushing and 30 minutes stagnation. If the average lead concentration from the 4 samples is >10 μg/l at more than 10% of sites, corrective measures should be implemented as well as additional sampling.

This new approach continues to be studied in Canada although the Province of Ontario has already implemented a regulatory standard of 10 μg/l for a 30 minutes stagnation sample. For Ontario utilities, compliance is based on the 90th percentile of the greater of two sequential 1 litre samples after the water has stood for 30–35 minutes (Ontario Ministry of the Environment, 2007). The issues raised with the US LCR will also apply in Canada, if Option 1 is implemented. Dilution effects can also affect the results of 30 minutes stagnation sampling in Option 2 and the averaging of potentially distorted data is questionable.

The guidance from Health Canada (2009) provides no definition of the term "optimisation" in the context of plumbosolvency control, other than references to compliance with the benchmarks derived from Options 1 and 2. Just as is the case in the US, there is scope in Canada for improving the way in which lead in drinking water is regulated and again it would be beneficial to achieve a better consensus on the definition of the term "optimisation" as it applies to plumbosolvency control in practical terms. These issues are the focus of this research project.

1.4 OPTIMISATION OF PLUMBOSOLVENCY CONTROL IN THE UNITED KINGDOM AND THE USE OF COMPUTATIONAL MODELLING TECHNIQUES

With further new European standards for lead in mind (25 μg/l from December 2003, 10 μg/l from December 2013), the Drinking Water Inspectorate (DWI) issued Information Letters (DWI, 2000, 2001) that required the water companies in England and Wales to optimise treatment measures to reduce plumbosolvency, including the dosing of orthophosphate. Optimisation of plumbosolvency control treatment in England and Wales was required if >5% of at least 100 random daytime (RDT) samples had exceeded 10 μg/l in the water supply zone(s) supplied by a treatment works and in consequence 95% of public water supplies are now dosed with orthophosphate for plumbosolvency control.

Optimisation was defined as the best practical reduction in lead concentrations and was taken to mean maintaining an optimum orthophosphate dose throughout a water supply system, within an optimum pH range. If pH and alkalinity control were proposed as the only treatment measures, water companies had to demonstrate that an optimum dose of orthophosphate could not achieve a further significant reduction in lead concentrations; in practice, none did so. References were also made to increasing orthophosphate doses to overcome organic colour and the need to minimise corrosion of iron mains within distribution systems because of its relevance to particulate lead.

The Information Letters implemented an optimisation framework and the precise definition of an optimum orthophosphate dose was necessarily vague. It was indicated that the optimum dose of orthophosphate could be determined from laboratory tests, from full scale or pilot scale trials, by practical experience, from solubility or computational models, or if an increase in orthophosphate dose produced no further worthwhile improvement. It was also stated that a water supply system could be considered optimised if a sufficient number of RDT samples had been taken and less than 2% samples exceeded 10 μg/l. This numeric criterion was adopted by most water companies as their target for optimisation, despite

DWI's initial preference for optimisation to be demonstrated by fixed point stagnation sampling. In practice, stagnation sampling from houses was found to be difficult to sustain, and the use of lead pipe test rigs was only partially successful.

The optimisation of orthophosphate dosing for plumbosolvency control can essentially be achieved in one of two ways:

- Step changes in dose until optimum reductions in lead have been demonstrated by lead pipe-rigs or in-situ lead pipes at consumers' houses, both likely to involve a stagnation sampling method; however, it is important to appreciate that old lead pipes can take up to two or three years to equilibrate with a new orthophosphate dose (IWA, 2010);
- Laboratory plumbosolvency testing coupled with computational compliance modelling to quickly determine the likely optimum dose, which should then be confirmed (and adjusted if necessary) by routine monitoring of in-situ lead pipes at consumers' houses; this approach can minimise the number of iterative changes to water treatment conditions and can save both time and money.

The success of this second approach has been reported in several published case studies (Hayes *et al.*, 2006, 2008; IWA, 2010) and was the motivation for this "proof-of-concept" project. Over 100 water supply systems have been modelled, involving 30% of the UK's water companies.

Importantly, the DWI followed up the progress being made by water companies with technical audits. Once concluded, optimisation schemes were reported formally. All optimisation schemes were subject to legally binding agreements with the Secretary of State. Northern Ireland and Scotland have separate administrative arrangements but these are closely aligned to England and Wales and broadly similar requirements for the optimisation of plumbosolvency control were applied.

The concentration of orthophosphate used in the UK varies from 0.5 to 2.0 mg/l (P), most typically between 1.0 and 1.5 mg/l (P), and is generally two or even three times higher than that dosed in the US (IWA, 2010). Doses are water supply system specific, and are determined by both water quality and the extent of occurrence of houses with lead pipes. Higher doses have been used in supplies that have an elevated organic content, particularly where humic substances are involved (Cardew, 2009). No problems have been reported in relation to bacterial growths within distribution networks and the impact on sewage effluent is considered slight (IWA, 2010).

The recent (2009 and 2010) compliance position in England and Wales for lead based on 25,641 RDT samples, 99.0% with 10 μg/l and 99.8% with 25 μg/l, is generally regarded as highly satisfactory. This compares to 80.4% compliance with 10 μg/l and 91.6% compliance with 25 μg/l over the period 1989 to 1994, before orthophosphate dosing became widespread, based on a total of 326,554 RDT samples (DWI, 1992, 1996, 2011).

In case studies that used computational modelling (Hayes *et al.*, 2006) the predicted failure to achieve 10 μg/l was found to be due to long lead pipes greater than 40 or 50 metres (depending on the zone involved). This suggested that additional selective lead pipe replacement that focused on long lead pipes could achieve even higher compliance, perhaps 99.8% (Hayes & Hydes, 2012). Recent guidance from the DWI (2010) requires water companies to maintain the optimisation of corrosion control treatment measures and to work with local authorities in the identification of any "hot-spots" where lead pipe replacement could be warranted.

There are generally no concerns in the UK about brass or the galvanic corrosion of leaded solder as being significant sources of lead in drinking water (Hayes & Hydes, 2012), although there have been a very small number of localised exceptions concerning brass manifolds (Hepple, 2008) and the illegal use of leaded solder (SCIEH, 2000). If lead leaching from such sources is not insignificant then it would appear that it is suppressed by optimised orthophosphate dosing.

1.5 PROJECT OUTLINE, OBJECTIVES AND BENEFITS

(a) Project outline

In the UK, computational modelling has been used extensively and very successfully in the optimisation of corrosion control in over 100 water supply systems. A single pipe model (with lead and non-lead components) predicts lead concentrations in drinking water as a function of the water's plumbosolvency, pipe-work characteristics and assumed water use. The single pipe model is used within a probability framework to predict lead concentrations across an entire water supply system (i.e. City or Town) and is coupled to a sampling model to predict compliance with standards. The UK uses random daytime (RDT) sampling to assess compliance; predictions of RDT sampling results from the modelling system have been validated repeatedly by the actual RDT sampling undertaken by water companies (Hayes *et al.*, 2006, 2008).

There was considered to be scope to amend the computational modelling system to be able to predict the results of surveys based on stagnation sampling, including sequential sampling. Once validated, the modelling system could then be used in the low-cost optimisation of plumbosolvency control for drinking water systems in Canada and the US.

(b) Objectives

The over-all objective of the project was to demonstrate the feasibility of an amended modelling system being used in the future optimisation of plumbosolvency control in Canadian and US drinking water supply systems.

The supporting objectives were:

(1) To amend an existing computational modelling system to be able to predict the results of sequential stagnation sampling (6 hours and 30 minutes).
(2) To develop validation procedures that can utilise stagnation sampling results provided by water utilities.
(3) To refine calibration inputs that reflected Canadian and US circumstances.
(4) To investigate the extent of the distortion effects associated with the currently specified stagnation sampling methods.
(5) To investigate the significance of variable influences on sampling methods.
(6) To demonstrate how the modelling system can be used in risk assessment.
(7) To develop a low-cost approach for optimising the corrosion control of lead in drinking water in the Canadian and US contexts.

(c) Benefits

The project aimed to:

(1) Provide a deeper insight into the behavioural characteristics of the stagnation sampling methods used in the regulation of lead in drinking water.
(2) Offer Canadian and US water utilities a low-cost approach to the optimisation of corrosion control in their systems.
(3) Provide a platform for further research in connection with improving the control of lead in drinking water in Canada and the US.

Chapter 2

Description of the computational compliance models

2.1 INTRODUCTION

The computational models, which are described in more detail elsewhere (Van der Leer *et al.*, 2002), enable the most relevant features of a water supply zone to be incorporated in the prediction of zonal compliance with lead standards, as a function of both plumbosolvency (corrosivity of the water to lead) and the zone's physical characteristics. A zonal model simulates the emissions of lead at individual simulated houses, through time, across an entire water supply zone or area of supply. It uses a single pipe model to determine the lead emission profile at each simulated house, the characteristics of each simulated house being the outcome of the random ascription of variables, which follows the Monte Carlo method for establishing a probabilistic frame-work. Such profiles can then be investigated by a sampling model to predict regulatory compliance.

It is assumed that the only sources of lead in drinking water are lead service lines and any internal lead pipes within premises and that lead contamination is due solely to the process of dissolution. The model does not simulate lead releases from brass or galvanic corrosion, nor particulate lead releases; however, correction factors can be applied if necessary.

2.2 THE SINGLE PIPE MODEL

The single pipe model simulates the dissolution of lead into the water flowing through or stagnating in a coupled lead pipe and non-lead pipe, over a 24-hour period. The coupled pipes are broken down into a series of "elements" (Figure 2.1) and when assuming simple plug flow, each element is treated as a stirred tank, flow being simulated by passing the contents of each stirred tank to the next at a time interval of one second. The rate of lead dissolution is determined by reference to an exponential curve that declines towards equilibrium, as illustrated in Figure 2.2.

As M (the initial mass transfer rate which determines the initial slope of the dissolution curve) and E (the equilibrium concentration) reduce, the water is less plumbosolvent (less lead dissolves; curves A to C) and these factors can be determined by stagnation sampling at appropriate reference houses or by laboratory plumbosolvency testing. Curves A1 and A2 differ in shape as a consequence of the relationships between the 30 minutes stagnation and equilibrium concentrations, which vary for individual waters (Hayes, 2008). The exponential curve and the assumption of plug flow are both approximations, but they enable the computational demands of the model to be greatly reduced. Extensive research (Hayes, 2002; Van der Leer *et al.*, 2002) has demonstrated that these approximations are adequate when compared to the more scientifically exact diffusion model and the three dimensional simulation of turbulent flow.

As a guide, a moderately plumbosolvent water (e.g. a high alkalinity groundwater that has not been orthophosphate dosed, or a low alkalinity water that has been pH adjusted to 8.0–8.5 but not orthophosphate dosed) will often be described by $M = 0.1\ \mu g\ m^{-2}\ s^{-1}$ and $E = 150\ \mu g\ l^{-1}$. For 12 mm internal diameter lead pipes, the equilibrium concentration (E) is predicted (Hayes, 2002) to occur after about 8 hours water stagnation, consistent with the observations and predictions of others (Kuch & Wagner, 1983).

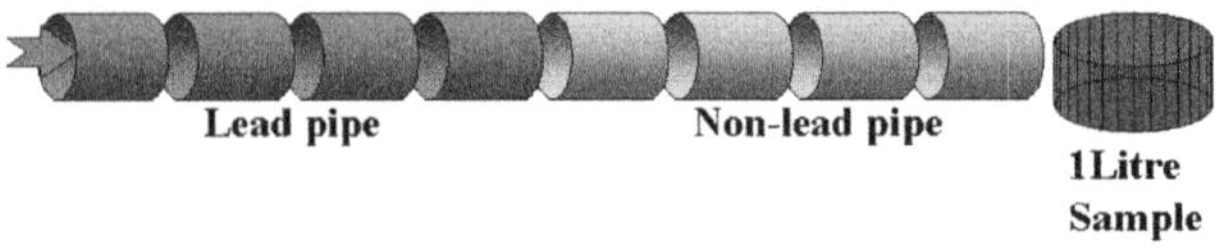

Figure 2.1 Configuration of the single pipe model.

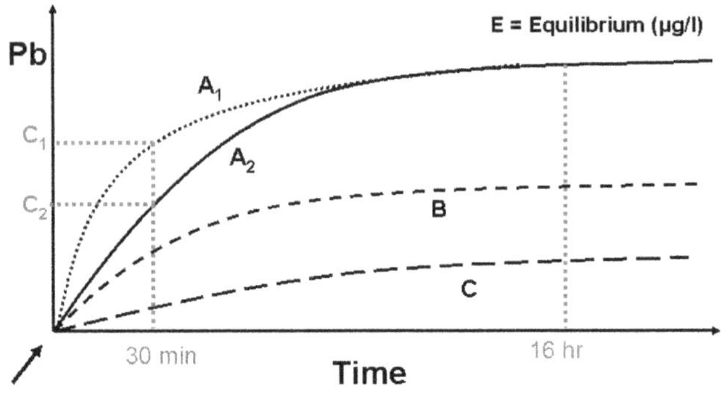

M = Initial mass transfer rate (µg/m²/sec), the initial slope. C_1 and C_2 are the 30MS concentrations of curves A_1 and A_2. B and C are curves for waters of lower plumbosolvency.

Figure 2.2 Lead concentrations as a function of stagnation time (from Hayes, 2009).

When the imaginary tap is closed (i.e. zero flow), the lead concentration increases over time as determined by *M* and *E*. When the imaginary tap is open, the concentration of lead in the emission from the pipe is either (i) a reflection of the steady state flushed condition (with lead concentrations normally below 1 µg/l unless the lead pipe is very long) or (ii) it is determined by previous zero flow (stagnation) conditions, as influenced by pipe geometry and the extent of the flow event. It can be appreciated that the simulation of such events in each stirred tank for every second of flow leads to many millions of calculations being performed for each simulated pipe.

2.3 THE ZONAL MODELLING FRAMEWORK

The zonal lead emission model is set up by the random ascription of a series of zonal characteristics, as derived from sets of agreed statistical distributions, and by the use of agreed variables and constants. The statistical distributions that are shown in Figure 2.3 have been used successfully in many zonal modelling studies (e.g. Hayes *et al.*, 2008); they were derived after extensive peer review and had the following features:

- the length of lead and non-lead pipes have a log-normal distribution, consistent with longer lengths occurring less frequently; the greater length of lead pipes relates to service connections often having to cross large house frontages, whereas non-lead premise plumbing is generally more compact;
- the volume used per day relates to an individual simulated house, the mean volume equating to the average water consumption of a house in the UK and assumed to flow through the simulated pipes;
- pattern A describes water usage in a house in which there is residency throughout the day when water is consumed (not during the night when residents are asleep);
- pattern B describes water usage in a house in which all residents are absent during "office hours" when no water is used;
- patterns A and B are applied for three and two water use frequencies, such that the weighting of A to B is 3 to 2, albeit with the water use frequencies within the two categories having an equal weighting.

For both the lead and non-lead pipes, 95% were normally assumed to have an internal diameter of 12 mm and 5% 18 mm, as related to UK conditions. Different diameter distributions can be used and applied separately to the lead and non-lead pipes. It is easy to change all of the above distributions if local survey data is available. The aim of this probabilistic Monte Carlo framework is to describe the huge variation that undoubtedly occurs in real water supply zones. If we can mimic this real-world variation then the model can be used for predictive purposes, as has been demonstrated by case studies (Hayes *et al.*, 2006, 2008).

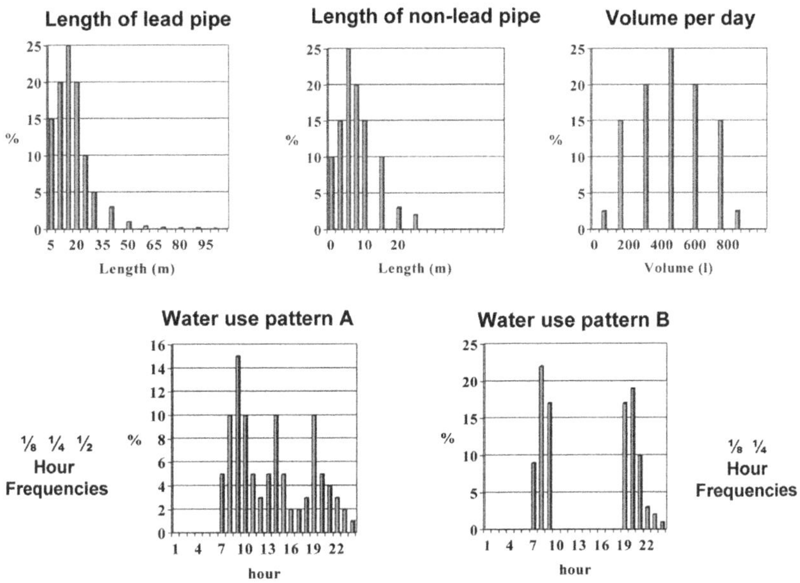

Figure 2.3 Statistical distributions used to set up the zonal model (from Hayes, 2009).

It should be appreciated that the average lead concentrations predicted by the model relate to a single plumbosolvency condition occurring in time, whether it is applied as a constant throughout an area or as a range. In consequence, the predicted results relate to an average condition over time. This is reasonable if the periods of time under consideration extend to multiples of a year, such that seasonal variation is accommodated.

The model, as used in numerous UK case studies, simply assumes that a single lead pipe supplies each simulated house, when a lead pipe is ascribed. However, the model does have an additional optional feature in which branched lead pipes can be simulated, to accommodate better multiple dwelling buildings (e.g. apartments), albeit case studies are presently limited.

The first step in the modelling process is to establish the probabilistic framework for the zone. Each time the program is run, a slightly different set of permutations will result from the range of calibration inputs. Variance in the model's outputs is no more than 1% in this respect. The sampling model then simulates the results of several sampling methods, which themselves exhibit variation that can be quantified from multiple simulations; examples are given in the following sections where appropriate.

2.4 SIMULATING SAMPLING

(a) Introduction

The sampling model can simulate several sampling methods and investigate compliance with regulatory standards:

(i) Average concentrations of lead at all or a random selection of simulated houses: this is relevant to the manner in which the WHO Guideline and EU standards are expressed;
(ii) Random daytime sampling of consumers' faucets across a water supply system: this has been used in risk assessment in the case studies for Cities A, B and C; it is the method used in regulatory compliance assessment in the UK (details below);
(iii) 6 hours stagnation sampling: of relevance to the US Lead Copper Rule and the case study for Cities A and C (details below);
(iv) 30 minutes stagnation sampling: of relevance to Health Canada's Guidelines, Ontario Provincial Regulations and the case study for City B (details below);

(b) Random daytime (RDT) sampling

In order to simulate a RDT survey, the specified number of simulated houses are selected at random and then a sampling time is selected at random between the hours of 09-00 and 17-00. The RDT sample is simulated by a stirred tank of one litre capacity as the outlet from the pipe. At the time of simulated sampling, the pattern of water use that has been applied to the simulated house is used to determine the immediately previous water–pipe contact position. It is routine to repeat the simulated survey, typically 100 or even 1000 times, in order to be able to understand possible variation. The result reported for the zone under investigation is the average survey result from all the surveys simulated (plus range and standard deviation, if required).

(c) 6 hours stagnation sampling

The percentage of simulated houses with lead pipes in the zonal model's probabilistic framework is set at 50% or the known percentage of samples that have been taken from houses with lead service lines (56.3% in the LCR surveys undertaken in City A). A specified (e.g.100) number of simulated houses are then selected at random for each simulated survey. In consequence, every survey will be made up from a different set of simulated houses. By simulating a large number of surveys, the average simulated survey condition will approximate to the average survey condition of actual LCR surveys. This simulation technique provides an insight into the extent of variation that might equally apply to actual surveys, bearing in mind that the Utility's sampling pool will change over time.

To simulate a LCR survey sample, it is assumed that the pipe-work is fully flushed before stagnation, in line with common practice in the US, and that the concentration of lead is zero. The first litre drawn after 6 hours' stagnation is then simulated at the imaginary faucet. Further one litre samples can then be simulated in sequence. Predicted lead concentrations after 6 hours stagnation are close to the selected equilibrium concentrations for lead, reflecting the dissolution curves generated by the model. For this reason, predicted lead concentrations after stagnation periods greater than 6 hours are little different to those predicted after 6 hours stagnation.

The results from a simulated survey are then assessed by calculating the 90th percentile concentration and comparing this to the Action Level of 15 μg/l; the percentage number of results exceeding 15 μg/l is also calculated.

(d) 30 minutes stagnation (30MS) sampling

The simulation of 30MS samples is the same as for 6 hours stagnation sampling, except that the stagnation time is 30 minutes and the standard used is 10 μg/l.

(e) Examples of model output

Examples are given in Appendix 2.

Chapter 3

Simulation of water flow in a pipe using computational fluid dynamics

3.1 INTRODUCTION

This section of the report compares laminar and plug flow along pipes of internal diameter 12 mm and 18 mm. Simulations were performed on a 30 m length of pipe which consisted of a 10 m length of lead pipe between two equivalent lengths of non-lead pipes. A series of one litre samples were taken and the proportion of each sample that started in the lead section of pipe was calculated.

The velocity of fluid flowing along a pipe varies with radius with a zero value at the pipe wall and a maximum on the pipe centre line. At low flow rates the flow will be laminar and the velocity at any radius, r, is proportional to $R^2 - r^2$ where R is the radius of the pipe. At higher flow rates the flow regime changes to turbulent and the velocity profile (how the velocity varies with radius) increases much more rapidly near the wall and changes very little over most of the pipe radius. An approximation to this velocity is plug flow where the velocity does not change at all with radius.

3.2 THE FLUID FLOW EQUATIONS AND THEIR COMPUTATIONAL SOLUTION

Fluid flow at the macro scale is governed by the conservation of mass and momentum. This conservation can be expressed through a set of differential equations. In general form the conservation of mass is expressed as:

$$\frac{\partial \rho}{\partial t} + \mathrm{div}(\rho \underline{u}) = S_m \quad (1)$$

In the above equation ρ represents the density of the fluid, $\underline{u}$ is its velocity vector and S_m accounts for any source of mass. For the conservation of momentum the equation for the ith Cartesian component of velocity is given by:

$$\frac{\partial(\rho u_i)}{\partial t} + \mathrm{div}(\rho \underline{u} u_i) = -\frac{\partial p}{\partial x_i} + \mathrm{div}\left[\mu\, \mathrm{grad}(u_i)\right] + S_i \quad (2)$$

In equation (2) the pressure is represented by the symbol p, μ represents the dynamic viscosity of the fluid and S_i includes all other sources into this momentum equation. The value of the subscript i takes values of 1 to 3 in order to provide equations for each of the components of the velocity vector.

Except in the simplest of cases an analytic solution to this equation set does not exist. Consequently computational techniques are employed to obtain an approximate solution. A number of techniques are available to convert the above equations into a form that can be resolved through computational methods. The method use in this work is the finite volume method (Versteeg & Malalasekera, 1995). The method converts each of the differential equations (1) and (2) into a matrix equation which can be computationally solved to obtain approximate values of the dependent variables, velocity and pressure, at a finite number of locations. The particular approach applied for the cases presented later employs a collocated discretisation method (Croft *et al.*, 1995) with a SIMPLEC (Dormal & Raithby, 1984) based method to couple the pressure and momentum fields and Rhie–Chow interpolation (Rhie & Chow, 1983) to avoid spurious pressure oscillations.

3.3 COMPARISON OF PLUG AND LAMINAR FLOW ALONG A STRAIGHT PIPE

Simulations were performed on both 12 mm and 18 mm diameter pipes. In all cases 30 m of pipe was simulated consisting of 10 m of lead pipe between two 10 m lengths of non-lead pipes.

For plug flow an inflow velocity, v_1, was specified to produce a flow rate of $q = 0.1$ litre/s. This velocity, in m/s, at the water inflow was calculated using the formula:

$$v_1 = \left(\frac{q}{1000}\right) \Big/ \left(\pi \frac{D^2}{4}\right) \tag{3}$$

In equation (3) D is the diameter of the pipe expressed in metres. A fixed gauge pressure of 0 Pa was applied to the outflow and a slip condition, no momentum sink, was applied at the walls of the pipe.

For cases involving laminar flow the outflow boundary was the same as the plug cases. To obtain the expected parabolic profile, associated with laminar flow, a non-slip condition was applied in the momentum equations. The boundary condition used at the inflow boundary was a fixed pressure condition. To generate a flow rate of Q m^3/s ($= q/1000$) along a pipe of length L, assuming laminar flow, it can be shown that the pressure drop along the pipe is given by the formula:

$$p_{\text{in}} - p_{\text{out}} = \frac{8\mu QL}{\pi R^4} = \frac{128\mu QL}{\pi D^4} \tag{4}$$

As a consequence of the zero pressure condition applied at the outflow boundary the inflow boundary pressure was set to the right hand side of (4).

In addition to solving for the velocity and pressure fields an extra variable, ϕ, is solved. The distribution of this variable is given by the following equation.

$$\frac{\partial \phi}{\partial t} - \operatorname{div}(\underline{u}\,\phi) = 1 \tag{5}$$

The value of this variable is set to zero at the outflow boundary and its values represent the time it will take fluid at any location to travel to the outflow boundary given the velocity field $\underline{u}$. Multiplying this variable by $4q/(\pi D^2)$, the flow rate divided by the pipe cross sectional area, produces a variable that can be used to indicate where each drawn litre starts from. The starting point of the water that contributes to the nth litre drawn from the pipes is identified by the positions where the value of $\Psi = 4q\phi/(\pi D^2)$ is between $n-1$ and n. Contour plots of Ψ for the four combinations of 12 mm or 18 mm diameter pipes and plug or laminar flow are contained in Figure 3.1 to 3.4. Each shade in those images represents a different litre.

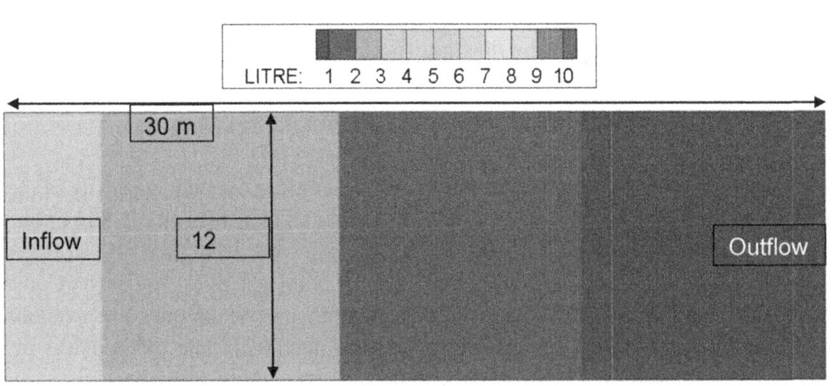

Figure 3.1 Starting location of each litre of water, 12 mm diameter pipe, plug flow.

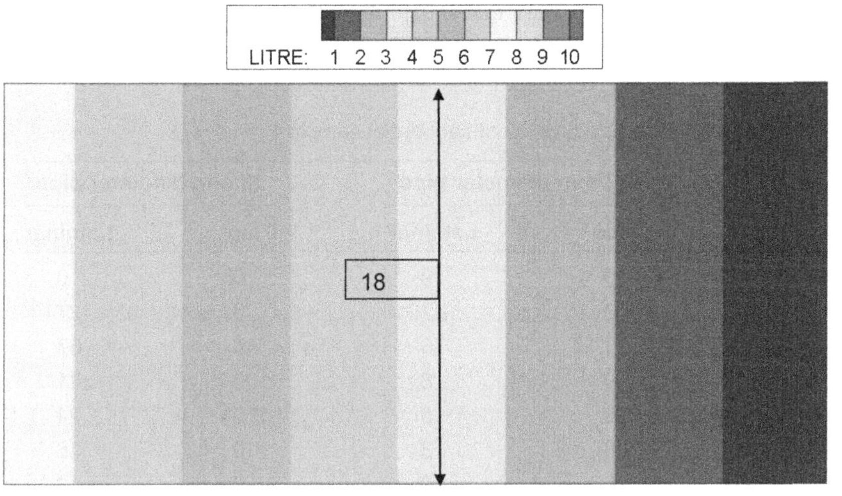

Figure 3.2 Starting location of each litre of water, 18 mm diameter, plug flow.

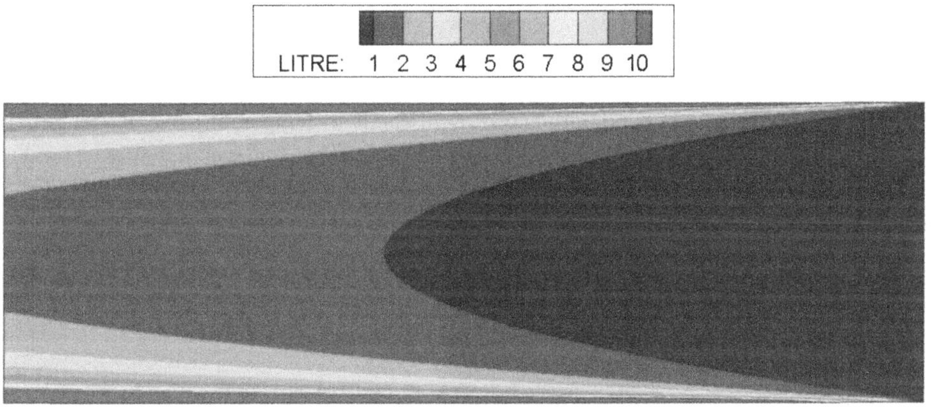

Figure 3.3 Starting location of each litre of water, 12 mm diameter pipe, laminar flow.

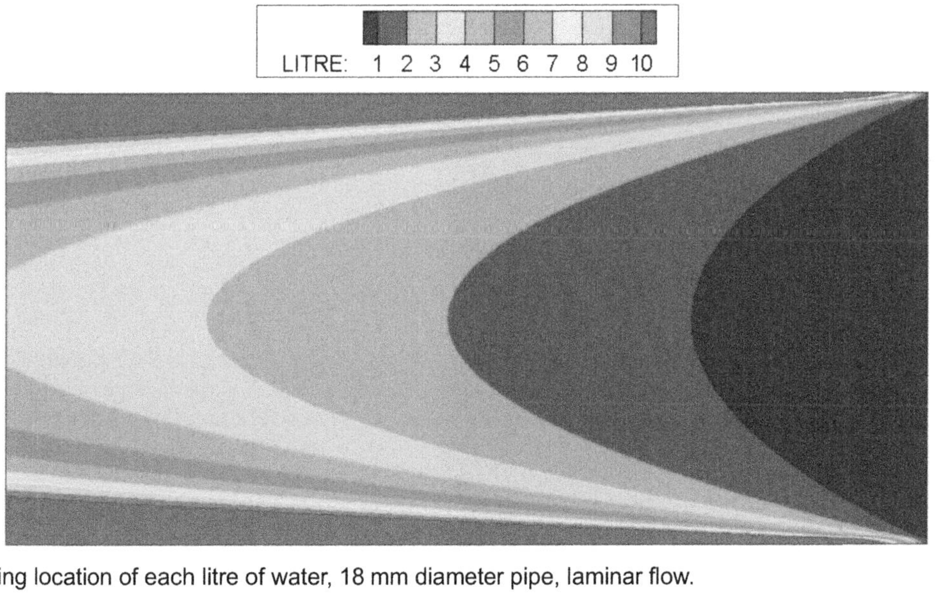

Figure 3.4 Starting location of each litre of water, 18 mm diameter pipe, laminar flow.

Table 3.1 below contains the percentage of water that started in the lead section of the pipes for each of the first 10 litre samples as calculated by the simulations. The skewing effect of laminar flow is clearly discernable.

Table 3.1 Volumetric profiles of sequential samples.

Litre	12 mm diameter pipe		18 mm diameter pipe	
	Plug	Laminar	Plug	Laminar
1	0	19	0	0
2	87	47	0	27
3	27	16	46	67
4	0	8	100	41
5	0	5	100	24
6	0	3	10	16
7	0	3	0	12
8	0	2	0	9
9	0	1	0	7
10	0	1	0	5

3.4 APPLICATION OF VOLUMETRIC PROFILES

The volumetric profiles shown in Table 3.1 are not influenced significantly by temperature over the range exhibited by seasonal variation in water supplies. The profiles apply to any plumbosolvency condition and can therefore be used in conjunction with predicted lead concentrations. The flow condition that applies can be determined by calculating the Reynolds number that relates to the pipe diameter and flow involved. The skewing effect is discussed further in Chapter 8.

Chapter 4

Calibration and validation

4.1 INTRODUCTION

The aim of building a computational model is to be able to use it for investigating complex processes or structures, to an extent far more than could ever be achieved experimentally. In the context of this project, the aim was to mimic the numerous permutations of the variables that determine lead emissions to drinking water across an entire water supply system, so that variables of interest could be investigated. The variables of greatest interest are those that affect the plumbosolvency of the water supplies, particularly factors *M* and *E*, such that optimisation of water treatment conditions can be pursued.

4.2 CALIBRATION

The success or failure of any model depends on how well it is calibrated. Some calibration inputs are more important than others and the accuracy of inputs will vary depending on data source. Figure 4.1 summarises the calibration inputs and their data sources of relevance to the zonal lead emission model; also shown are the methods of validation and model outputs.

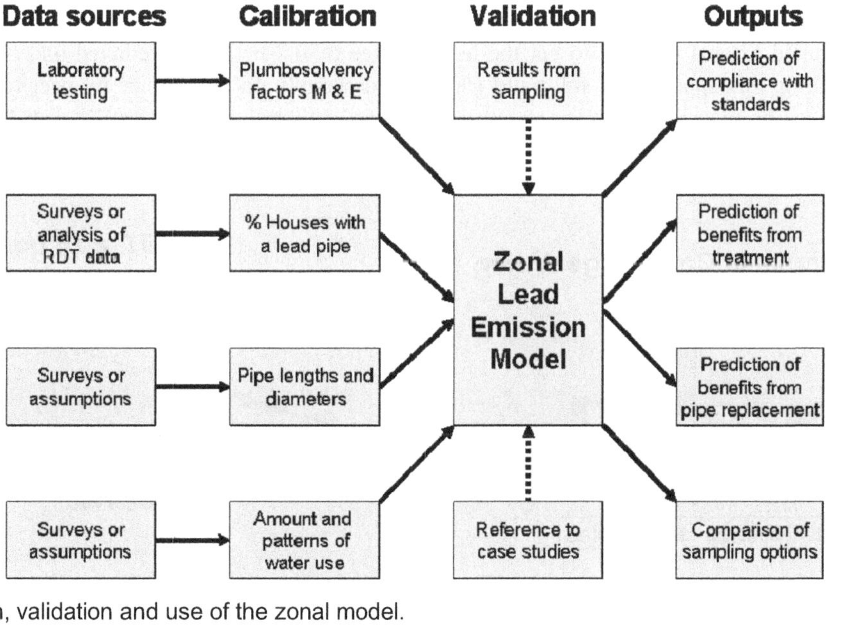

Figure 4.1 Calibration, validation and use of the zonal model.

In practice, calibration is never perfect and assumptions often have to be used, after appropriate peer review. This is illustrated by Figure 4.2, which summarises the calibration status of the three North American case studies, with two from the UK for comparison.

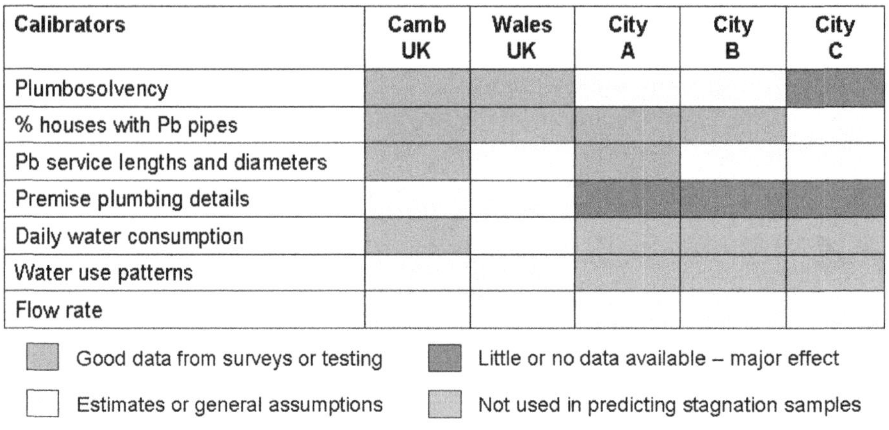

Figure 4.2 Calibration status of case studies.

For the two UK case studies, the most important calibration data were the plumbosolvency of the water, the percentage of houses with lead pipes within the water system, and daily water consumptions. Pipe details, water use patterns and flow rates were less critical and general assumptions were often used (Hayes *et al.*, 2006, 2008). For the North American case studies in Cities A, B, and C, the most important calibration data were the plumbosolvency of the water and the length and diameter distributions of the non-lead pipes used in premise plumbing. In these studies the focus of compliance assessment was either sequential 30 minutes or 6 hours stagnation – first draw sampling. For City C, the limited data available for model calibration constrained the scope of the case study.

4.3 VALIDATION

In the UK studies, validation of the model was achieved by comparing predicted RDT sampling results to those from actual RDT sampling by the water companies. The excellent validation achieved in the Cambridge and Wales case studies is demonstrated in Figures 4.3 and 4.4. By implication and very importantly, the simple assumptions and limitations of the zonal compliance model (Chapter 2) and the use of generalised assumptions in calibration, clearly do not impede the construction of models that can be used to fulfil their intended purpose. For the North American case studies in Cities A, B, and C, validation by independent data was not possible due to a lack of calibration data. Instead, the calibration data was adjusted (to a limited extent) in order to get the best "degree of fit" between predicted and observed survey results. Whilst not so compelling, this approach is still valid and depends largely on acceptance by peer review. In simple terms, the calibration data must be plausible. The least well characterised calibration data was the distribution of non-lead pipe lengths within premises which, somewhat ironically, exerts a major influence on regulatory compliance when it is based on stagnation sampling.

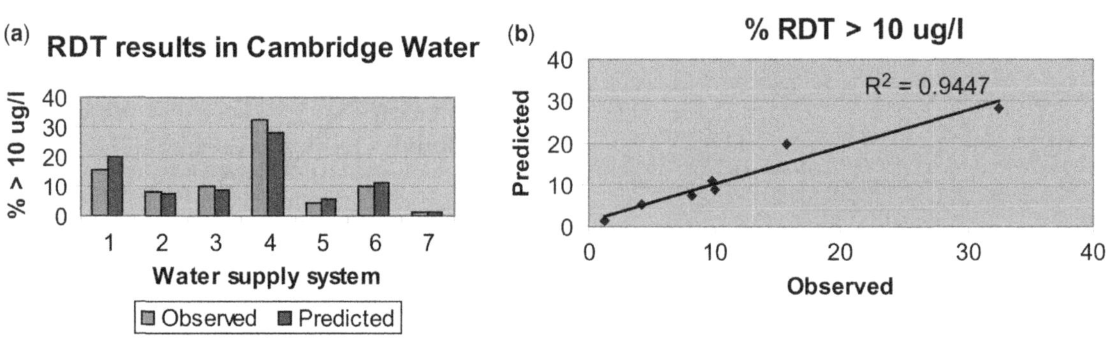

Figure 4.3 Validation of the model in the Cambridge Water case study (Hayes *et al.*, 2006). (a) Observed and predicted RDT results; (b) Correlation between the observed and predicted results.

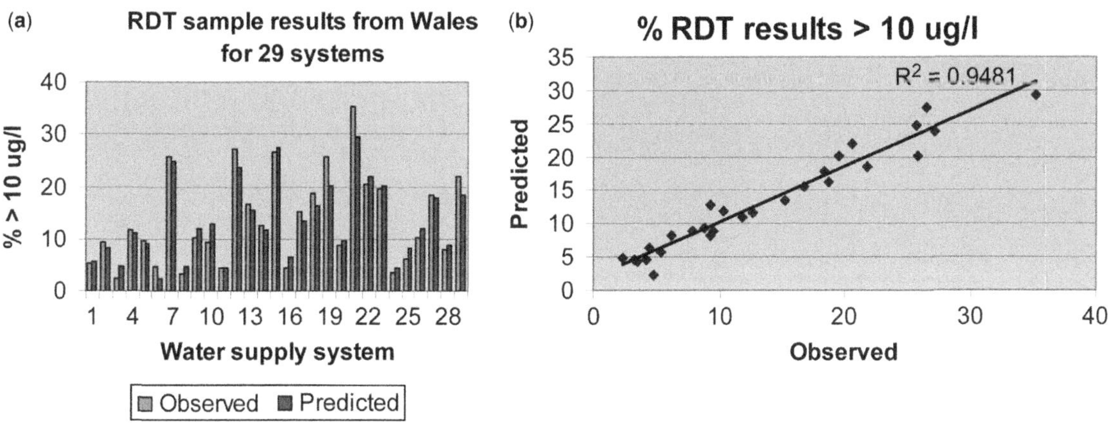

Figure 4.4 Validation of the model in the Wales case study (Hayes *et al.*, 2008). (a) Observed and predicted RDT results; (b) Correlation between the observed and predicted results.

Chapter 5

Case study: City A (US)

5.1 BACKGROUND

The water supply system serves a population of around 300,000 people with acknowledged high quality drinking water. The Utility has 74,000 customer connections of which 26,000 (35%) are lead service lines (LSL), including the privately owned side. Additionally, the Utility provides water supplies to neighbouring Utilities that serve a further 300,000 people. Water is abstracted from a series of reservoirs and treated by rapid sand filtration with conventional chemical treatment before pumping into distribution.

The treated water entering supply typically has a pH of 10.0, an alkalinity of 18 mg/l (as $CaCO_3$) and a total organic carbon content of 2.1 mg/l. The main anions (chloride, nitrate, sulphate) are low in concentration. The orthophosphate concentration after chemical treatment is very low at 40 µg/l.

Free chlorine entering supply is at about 1 mg/l but this attenuates to around 0.4 mg/l within the distribution network. Consistent with such conditions, the analysis of lead pipe deposits (by the US EPA, Cincinnati) revealed the presence of Pb(II) compounds; these are more soluble than the Pb(IV) compounds that can occur at higher free chlorine concentrations.

The iron content of the treated water entering supply is very low, at about 20 µg/l, but does increase to cause iron discolouration problems in some parts of the distribution network, particularly at the extremities where water stagnation times within old cast iron water mains can be more significant. Occasional "spikes" in iron can result in intermittent particulate lead problems at some premises.

In order to control the plumbosolvency of the water supplies to consumers, it has been long-standing practice to elevate pH in order to suppress lead solubility. The pH within the system is typically between 9.5 and 10.0. Corrosion inhibitors such as orthophosphate are not used. This approach to plumbosolvency control has succeeded in general terms, with 80 to 90% of 6+ hours stagnation samples being below the Action Limit of 15 µg/l over the period 2007 to 2011. However, the system was marginally non-compliant with the criteria for lead set by the US Lead Copper Rule in 7 out of 9 surveys over this period.

5.2 CALIBRATION AND USE OF THE LEAD EMISSION MODEL

(a) Lead pipe lengths and diameters

Comprehensive data provided by the Utility for full and partial lead services were combined. The data distributions used are as shown in Appendix 1 (Tables A1 and A2).

(b) Non-lead pipe lengths and diameters

Very little data were available for premise plumbing. The length distribution that gave the best fit between predicted and observed LCR survey results is shown in Appendix 1 (Table A3). This distribution is an amended version of the one used in UK studies. The diameters applied to non-lead pipes were the same as for lead pipes (Table A2). This data is only relevant to the situation where a lead pipe is coupled to a non-lead pipe. Non-lead service lines were quantified separately by reference to the percentage of lead service lines (56.3%) that were sampled in LCR surveys.

(c) Water consumptions and patterns of use

These aspects are not relevant when predicting the results of stagnation sampling but are used when predicting the results of random daytime (RDT) sampling. The data distributions used in UK modelling studies are as shown in Figure 2.3.

(d) Plumbosolvency factors

The initial mass transfer rate (M in $\mu g/m^2/s$) and the equilibrium concentration for lead (E in μg/l) can be derived from rapid laboratory plumbosolvency testing if Pb(II) chemistry dominates the corrosion scale within the lead pipes. However, no such data was available. Instead, reference was made to sequential sampling exercises by both the Utility and the State Health Authority, and to LCR survey results, in an attempt to pick out high lead concentrations that might be reasonably representative of equilibrium conditions (E).

The following could be discerned:

- (i) From the 9 LCR surveys over the period 2007 to 2011, the average of the 3rd highest LCR survey results was 69.4 μg/l; the 1st and 2nd were ignored as they were often much higher and were assumed to reflect particulate interference.
- (ii) The Utility's sequential sampling exercise at one residential property had a total lead concentration of 100 μg/l and a dissolved lead concentration of 68 μg/l associated with the lead pipe adjacent to the faucet.
- (iii) The average highest lead was 84.6 μg/l, before partial lead service line replacement, at 8 houses sequentially sampled in the summer of 2010 by the State Health Authority (representativeness is not known); an adjustment of $\times$ 0.75 needs to be made to reflect an average annual condition, on the basis that lead concentrations tend to be about twice that in the summer compared to the winter (IWA, 2010), calculated to be 69 μg/l.
- (iv) On the above basis, E was estimated to be 70 μg/l and on the basis of the common relationship between $M = 0.1$ and $E = 150$, a value for M of 0.047 was calculated.

(e) Other model inputs

In all cases, a zone with 10,000 houses was simulated which is adequate for modelling purposes. For the simulation of LCR surveys, the flow rate for simulated periods of flow was set at 0.1 l/s in conjunction with the plug flow model. Each LCR survey simulated assumed 56.3% houses had a lead service pipe, based on the Utility's LCR data. To simulate LCR results, 100 samples were selected from simulated houses at random and were repeated 1000 times to get an average predicted result and the extent of variation around the average.

(f) Uncertainties

The main uncertainties were the plumbosolvency factors that were derived indirectly as a best estimate and the length distribution for non-lead premise pipes which had to be assumed. Independent validation of the predicted results was not possible. However, the model was calibrated to achieve a good fit with the observed LCR survey results, as demonstrated in Section 5.3.

(g) Premise plumbing

It was assumed in the UK case studies that no lead arose from non-lead pipes but this was at odds with the LCR data provided by the Utility in which 3.4% of the total number (1023) of samples, from the 9 surveys combined, that exceeded 15 μg/l were from houses with a copper service pipe. In comparison, 13.7% of the total number of samples, from the 9 surveys combined, that exceeded 15 μg/l were from houses with a partial or full lead service pipe. On this basis, 80% of exceedences would be explained by lead service pipes.

Further analysis of the LCR survey results revealed that 43.7% of houses had a copper service line with no lead. Whereas 72.2% of these recorded a lead concentration of <5 μg/l, lead was recorded in the remainder to varying extents. This has enabled a correction factor to be determined, the profile of which is shown in Table A4 of Appendix 1.

The simple assumption has been made that no lead arises from a copper service line and that the observed elevated lead concentrations arise from premise plumbing. This is reasonable as it is the premise plumbing that contains the numerous brass components, brass piping in some cases and legacy leaded solder, not the copper service line. It is also reasonable to assume that the lead arising from premise plumbing similarly occurs at houses with a lead service line. The correction for premise plumbing has therefore been applied to all houses, not just those with copper service lines. The correction has been made by random ascription within the probability framework of the model, using a data distribution that follows Table A4 of Appendix 1.

In studies on houses in Montreal (Cartier *et al.*, 2011) 4% of premise plumbing included a lead pipe, whereas in Cambridge (UK) 5% of premise plumbing was found to include a lead pipe. If these studies are extrapolated to City A, it is possible that between 50 and 63% of observed LCR exceedances from houses with copper service lines might be explained by lead pipes within premises. The possibility of lead pipes occurring in premise plumbing is the reason why the data distribution for non-lead pipes was set with 5% having zero length, so that the simulated lead pipes connect directly to the faucet in this minority case (Table A4 in Appendix 1).

The LCR survey data was also assessed to see if there were any signs of galvanic corrosion at copper-lead service line joints, by reference to atypical "spikes" in the lead results. The total number of copper-lead cases from all the LCR surveys combined was 27. Of these, 14 (51.85%) had lead <5 μg/l, 7 (25.93%) had lead between 5 and 10 μg/l, 3 (11.11%) had lead between 10.1 and 15 μg/l, and 3 (11.11%) had lead >15 μg/l. This profile, based on the 27 cases of copper-lead service lines, is fairly similar to the profile of all lead concentrations observed from all the LCR surveys, and it therefore appears that galvanic corrosion at the copper-lead service line joint is not a major issue.

5.3 RESULTS

(a) Matching predicted to observed LCR survey results

The calibration of the zonal compliance model was an iterative process that involved matching predicted compliance LCR data to that observed by the Utility, and it is important to put this into perspective:

(i) the data from the Utility on LSL occurrence, lengths and diameters were very comprehensive and was used without modification;
(ii) the quantification of the plumbosolvency factors *M* and *E* was limited to the extrapolation of survey data from LCR surveys and two sequential sampling exercises; although this quantification must be regarded as preliminary, the factors determined do compare fairly well with data from the UK; once determined, the values of *M* (0.047) and *E* (70) were used without modification;
(iii) the correction factors for premise plumbing were established by inspection of the Utility's LCR survey data (9 surveys in total with over 1000 results available) and were used without modification;
(iv) the calibration data for the length and diameter of non-lead pipes (premise plumbing) was based on simple assumptions which were modified in order to achieve the best match between predicted and observed LCR compliance data.

The resulting match between predicted and observed LCR survey results is shown in Table 5.1, based on the first litre sampled after stagnation. The predicted data in Table 5.1 are the average from three simulations, with variance of 3.5% for the average 90th percentiles and 4.2% for the average percentages of samples exceeding 15 μg/l. The predicted ranges that are shown relate to the three simulations taken together.

Table 5.1 Predicted and observed LCR survey results*.

Average 90th percentile concentration and range (μg/l)	Average percentage of samples exceeding 15 μg/l and range
Predicted: 20.5 (6.3 to 37.8)	Predicted: 14.6 (4.0 to 27.0)
Observed: 20.1 (13.6 to 30.0)	Observed: 16.2 (8.0 to 29.4)

*Predicted average 90th percentiles from 1000 surveys, each of 100 samples.

The predicted and observed data match closely and indicate that the model was adequately calibrated, providing a good starting point for modelling investigations.

(b) Orthophosphate dosing scenarios

Dosing scenarios were then investigated by reducing the values of *M* and *E* that define plumbosolvency. As these values were reduced, an equivalent reduction factor was applied to the premise plumbing correction. The results are shown in Table 5.2 for four sequential litre samples, using the plug flow model, and are compared to the present non-orthophosphate dosed condition.

Table 5.2 Predicted LCR compliance for orthophosphate dosing scenarios*.

	Plumbosolvency		Predicted average 90th percentile concentrations (μg/l)			
Scenario	***M***	***E***	**L1**	**L2**	**L3**	**L4**
Current–no dosing	0.047	70.0	20.5	63.8	67.0	59.1
o-PO_4 (A)	0.020	30.0	10.1	27.8	28.8	25.9
o-PO_4 (B)	0.015	22.5	9.6	20.8	21.6	18.5
o-PO_4 (C)	0.010	15.0	8.7	14.2	14.5	13.1

*Predicted average 90th percentiles from 1000 surveys, each of 100 samples.
M = initial mass transfer rate (μg/m^2/s); E = equilibrium concentration (μg/l).
L1 to L4 are the sequential litre samples. A, B and C represent different orthophosphate doses.

The orthophosphate dosing scenario A with $M = 0.02$ and $E = 30$ is typical of many optimised systems in the UK and should be achievable with a relatively low orthophosphate dose. It is predicted that LCR compliance would be achieved, based on the first litre drawn after stagnation, but not if compliance was based on further sequential samples. This is also the case with B when $M = 0.015$ and $E = 22.5$. To achieve LCR compliance based on sequential sampling would require $M = 0.01$ and $E = 15$ (scenario C), likely to be at the extreme range of orthophosphate's ability to suppress lead solvency. Laboratory plumbosolvency testing will be necessary to confirm the orthophosphate dosing response of the water in City A (i.e. the doses required for scenarios A, B and C), and would help to confirm these compliance predictions (which would be subject to operational feasibility).

(c) Risk assessment

On the basis that the model was adequately calibrated, it was possible to predict the results of random daytime sampling for the water supply system and to predict the percentage of samples that exceeded the WHO guideline value of 10 μg/l. The percentage of failing RDT samples equates to the percentage of population at risk on this basis.

The simulations assumed that 35.5% houses had a lead service pipe, based on the Utility's data (not to be confused with the percentage of LSLs included in LCR compliance surveys). An allowance was then made for other lead contributions from premise plumbing by applying a simple correction factor of $\times$ 1.25. This correction factor reflects the ratio of LCR results $>$15 μg/l for houses with copper services (20%) and houses with lead services (80%).

For the present situation and the three orthophosphate dosing scenarios, the predicted average percentage of simulated RDT samples exceeding 10 μg/l was determined from 1000 surveys each with 100 samples. The results, amended for premise plumbing, are shown in Table 5.3 and demonstrate that risk reduction is proportional to orthophosphate dose (or other treatment measure).

Table 5.3 Predicted RDT survey results*.

	Plumbosolvency		Simulated RDT samples
Scenario	***M***	***E***	**% >10 μg/l**
Current–no dosing	0.047	70.0	5.20
o-PO_4 (A)	0.020	30.0	0.59
o-PO_4 (B)	0.015	22.5	0.19
o-PO_4 (C)	0.010	15.0	0.01

*Predicted RDT results from 1000 surveys, each of 100 samples.
M = initial mass transfer rate (μg/m^2/s); E = equilibrium concentration (μg/l).
A, B and C represent different orthophosphate doses.
Results amended for premise plumbing.

It is interesting to compare these predicted results to what has been achieved in the UK by optimised orthophosphate dosing. At anticipated values of *M* and *E* in the region (i.e. both above and below) of 0.02 and 30, respectively, non-compliance across England and Wales is now down to 1% and in several regions it is down to 0.5%.

5.4 DISCUSSION

Analysis of corrosion deposits from 6 lead pipes exhumed from City A indicated a dominance of Pb(II) compounds, which are more soluble than Pb(IV) compounds. This is consistent with the moderately high lead concentrations found in both LCR surveys and sequential sampling exercises. Despite pH elevation to 10.0 at the treatment works (which does not normally fall below 9.5 in the distribution network), 7 of the 9 surveys failed the LCR for lead over the period 2007 to 2011. The dominance of Pb(II) compounds also means that rapid laboratory plumbosolvency testing can be used in the quantification of *M* and *E*, and that the supply system should be amenable to orthophosphate treatment (albeit the transition characteristics of lead corrosion deposits under high pH conditions would have to be determined).

Iron (red water) discolouration is a problem in some parts of the distribution network, particularly with dead-end mains and will likely influence the formation of particulate lead, resulting in occasional spikes of high lead, beyond that generally explained by dissolution alone. For this reason, the highest and second highest LCR results were ignored when attempting to gauge the lead solubility equilibrium concentration (*E*). Regardless, the LCR results indicate an underlying lead dissolution problem, despite the distorting effect of water stood in non-lead pipes adjacent to the faucet being sampled. In this respect, the sequential sampling exercises undertaken by the Utility and by the State Health Authority confirmed that lead concentrations in the first litre drawn after 6+ hours stagnation can be significantly lower than subsequent samples where the water has been in contact with lead piping. This is discussed further in Section 8.

A further likely influence on plumbosolvency is the organic content of the water, which averaged 2.1 mg/l total organic carbon (TOC). This could explain why pH elevation has not been wholly effective. The limitations of pH elevation (generally not above 9.0) were a major reason for the UK Government insisting on the dosing of orthophosphate. There are numerous examples in the UK and elsewhere in Europe of the limitations of pH elevation alone, albeit at pH values up to around 9.0, slightly lower than in City A.

The level of compliance with 10 μg/l in England and Wales has reached 99.5% in 4 out of 7 regions, at least in one year over the period 2005 to 2010 and it has been speculated (Hayes & Hydes, 2012) that further fine-tuning of corrosion control treatment coupled with selective lead pipe removal might deliver 99.8% compliance with 10 μg/l (on the basis of random daytime sampling). This could imply that in the UK: (1) sources of lead that exceed 10 μg/l other than lead pipes could be limited to around 0.2% of houses and appear unlikely to exceed 0.5%, and/or (2) sources of lead other than lead pipes could be potentially more significant but have been suppressed by optimised orthophosphate dosing. This issue could be relevant to City A where orthophosphate is not dosed presently and where high lead concentrations have occasionally been found in LCR surveys at properties without a lead service line. The anticipation is that optimised orthophosphate dosing in City A would suppress lead dissolution from both lead service lines and premise plumbing.

With this in mind, the three orthophosphate dosing scenarios were modelled. The values of *M* (0.02) and *E* (30) are readily achievable, based on UK experience, and demonstrated a significant further reduction in lead concentrations that would (if achieved in practice) comply with the LCR. To determine the likely dose of orthophosphate, it would be a fairly simple matter to undertake laboratory plumbosolvency testing with results obtainable within a month. Any organic influences should be captured by such testing.

5.5 CONCLUSIONS

(1) Computational modelling has been demonstrated to offer further insights into plumbosolvency and its control in City A.
(2) Plumbosolvency testing to confirm orthophosphate dose responses at different pH values could be considered.
(3) The Utility could also consider obtaining basic details of premise plumbing when the opportunities arise.
(4) If the Utility was minded to dose orthophosphate for further plumbosolvency control, a step-wise dosing strategy could be devised that accommodated the potential future use of sequential samples in the optimisation of corrosion control, as is already the case in Canada.
(5) The operational feasibility of a transition to orthophosphate dosing would first need to be confirmed.

Chapter 6

Case study: City B (CA)

6.1 BACKGROUND

The water supply system serves a population of around 800,000 people with an acknowledged high quality of drinking water. The Utility has 210,000 customer connections of with 14% are believed to have lead service lines (LSL). Two water treatment plants treat surface derived water and feed 11 pressure zones. Analysis of corrosion deposits from lead pipes exhumed from City B has indicated a dominance of Pb(II) compounds, which are more soluble than Pb(IV) compounds. The treated water entering City B has a pH of 9.2 to 9.4. This is sustained generally within the distribution network although pH can drop in rare cases to around 8.5. This general pH stability is consistent with the alkalinity of the water which averages 35 mg/l (as $CaCO_3$). The average dissolved organic carbon of the treated water is 3 mg/l, indicative that organics could influence plumbosolvency. The typical chlorine residual in City B is 1.7 mg/l, mostly as monochloramine, consistent with the occurrence of Pb(II) corrosion deposits.

In order to control the plumbosolvency of the water supplies to consumers, it has been the Utility's long-standing practice to elevate pH in order to suppress lead solubility. This approach has been successful in meeting regulatory compliance with Provincial standards. Corrosion inhibitors such as orthophosphate have not been used and whilst the possibility of further suppressing lead concentrations is appreciated by the Utility, there are concerns about the potential for biofilm development in the distribution network and adverse environmental impact.

6.2 CALIBRATION AND USE OF THE LEAD EMISSION MODEL

(a) Lead service pipe lengths and diameters

For lead service line (LSL) lengths, no data were available other than an estimate of a typical length of 13.5 m. The data distribution that was used for lead pipe lengths is shown in Appendix 1 (Table B1) and was derived from: (a) lead service line lengths observed in Montreal (Cartier *et al.*, 2011) and (b) minor adjustments to help fit predicted 30MS survey results to those observed. It equates to an average length of 12.4 m. For LSL diameters, the consensus from supply network engineers is shown in Table B2 of Appendix 1.

(b) Non-lead pipe lengths and diameters

No data were available for premise plumbing other than estimates of the pipe materials involved: >90% copper, 2% galvanised iron, 8% plastic. The distribution that gave the best fit between predicted and observed 30MS survey results is shown in Appendix 1 (Table B3) and was initially derived from the studies in Montreal. Generally smaller diameters were assumed for premise plumbing, broadly similar to Montreal, as shown in Table B2 of Appendix 1.

(c) Water consumptions and patterns of use

These aspects are not relevant when predicting the results of stagnation sampling but are used when predicting the results of random daytime (RDT) sampling. The data distributions used in UK studies are shown in Figure 2.3.

(d) Plumbosolvency factors

The initial mass transfer rate (M in $\mu g/m^2/s$) and the equilibrium concentration for lead (E in μg/l) can be derived from rapid laboratory plumbosolvency testing if Pb(II) chemistry dominates the corrosion scale within the lead pipes. No such data was available. Instead, reference was made to the 6+ hrs and 30MS sequential sampling exercises by City B. The initial estimate of E was 25 μg/l and was refined to 31 μg/l by calibration trials. This equilibrium concentration would be expected (from numerous earlier relationships found in UK studies) to be associated with a value for M of just over 0.02. Calibration trials refined this to 0.026.

(e) Other model inputs

In all cases, a zone with 10,000 houses was simulated which is adequate for modelling purposes. For the simulation of 30MS, 6 hrs stagnation and RDT surveys, the flow rate for simulated periods of flow was set at 0.1 l/s in conjunction with the plug flow model. Each 30MS survey simulated assumed 95% houses had a lead service pipe, consistent with earlier surveys by the Utility. Each 6 hrs stagnation survey simulated assumed 50% houses had a lead service pipe, to reflect Health Canada's guidelines (2009). Each RDT sampling survey that was simulated assumed 14% of houses had a lead service pipe, based on the Utility's estimate. To simulate survey results, 100 samples were selected from simulated houses at random and this was repeated 1000 times in order to determine an average predicted result and variation.

(f) Uncertainties

The main uncertainties were the plumbosolvency factors that were derived indirectly as a best estimate and the length distribution for non-lead premise pipes which had to be assumed. Independent validation of the predicted results was not possible. However, the model was calibrated to achieve a good fit with the observed 30MS survey results, as demonstrated in Section 6.3.

(g) Premise plumbing

A correction for lead release from premise plumbing was not considered necessary because: (i) the stagnation time was much shorter than used in City A; (ii) the Utility had data that lead contributions from premise plumbing were insignificant; (iii) brass pipes have never been used in City B; and (iv) survey data had very few atypically high lead values.

6.3 RESULTS

(a) Predicted and observed 30MS survey results

The match between predicted and observed 30MS survey results is shown in Table 6.1, based on the first four litres sampled after flushing and stagnation.

This shows that the model had been calibrated closely to observed survey results when $M = 0.026$ and $E = 31$ and that the model could be used for investigative purposes. The slight reductions in M to 0.020 and E to 30 resulted in a very substantial change in compliance. This is because 30MS values of 10 μg/l cannot be exceeded by the lead dissolution curve generated by these latter values (the highest predicted 30MS concentration is 9.69 μg/l). The higher values of M (0.030) and E (45) indicate substantially worse compliance and illustrate the sensitivity of the model.

(b) Predicted results for sequential sampling after 6 hrs stagnation

The simulated results for sequential sampling after 6 hrs stagnation are shown in Tables 6.2 and 6.3. With $M = 0.026$ and $E = 31$, compliance was achieved with Health Canada's guidelines for Tier 1 sampling but not for Tier 2 sampling if judged on the same 90th percentile basis. The difference is the dilution effect of water stood in non-lead pipe-work. A reduction to $M = 0.010$ and E of 15 will be required for Tier 2 compliance, likely to be at the limit of what plumbosolvency control treatment can achieve.

Table 6.1 Predicted and observed 30MS survey results.

	Average % 30MS samples >10 µg/l			
	1st Litre	**2nd Litre**	**3rd Litre**	**4th Litre**
Predicted with $M = 0.026$ & $E = 31$ (range in brackets)	0.21 (0.00 to 3.00)	1.92 (0.00 to 7.00)	7.79 (1.00 to 18.00)	15.23 (4.00 to 25.00)
Observed over 8 surveys from 2008 to 2011 (range in brackets)	1.33 (0.00 to 2.56)	2.36 (0.00 to 5.98)	8.55 (2.42 to 16.24)	11.95 (1.92 to 26.50)
Predicted with $M = 0.020$ & $E = 30$* (range in brackets)	0.00 (0.00 to 0.00)	0.00 (0.00 to 0.00)	0.00 (0.00 to 0.00)	0.00 (0.00 to 0.00)
Predicted with $M = 0.030$ & $E = 45$ (range in brackets)	2.70 (0.00 to 9.00)	15.22 (6.00 to 26.00)	50.12 (31.00 to 65.00)	44.53 (31.00 to 58.00)

*Typical optimised condition in the UK.

Table 6.2 Predicted LCR compliance.

Plumbosolvency		Predicted average 90th percentile concentrations (µg/l)			
M	*E*	**1st Litre**	**2nd Litre**	**3rd Litre**	**4th Litre**
0.026	31	0.09	20.20	30.54	30.62
0.020	30	0.12	19.54	29.09	29.08
0.010	15	0.01	9.22	14.54	14.57

$M =$ Initial mass transfer rate (µg/m^2/s); $E =$ equilibrium concentration (µg/l).

Table 6.3 Predicted Tier 1 (1st litre) and Tier 2 (2nd, 3rd & 4th litres) compliance*.

Plumbosolvency		Predicted average percentage samples >15 µg/l after 6 hours stagnation			
M	*E*	**1st Litre**	**2nd Litre**	**3rd Litre**	**4th Litre**
0.026	31	1.41	13.62	34.90	31.62
0.020	30	1.45	14.01	35.11	30.87
0.010	15	0.00	0.00	0.00	0.00

*Corrective action required if more than 10% sites exceed 15 µg/l;

The results in Tables 6.2 and 6.3 for $M = 0.026$ and $E = 31$ can be compared to the typical optimised orthophosphate dosing scenario of $M = 0.02$ and $E = 30$ in the UK. It is apparent that the case for supplementary orthophosphate dosing in City B is fairly weak. However, the discrepancies between compliance based on 30MS and 6 hrs stagnation sampling can be noted, by reference to $M = 0.020$ and $E = 30$.

In an earlier investigation, the Utility compared the lead concentrations in sequential samples after stagnation for 30 minutes and 6 hours at 22 homes. On average, for each litre sampled sequentially, the lead concentrations after 6 hours stagnation were 1.1 to 1.4 times higher than the lead concentrations after 30 minutes stagnation, much closer than expected. The ratio between 30 minutes and 16 hours stagnation is normally in the range 3 to 5 but can be as low as 1.8 and as high as 15.2 for waters without orthophosphate (Hayes, 2008); on the basis of the lead dissolution curves generated by the model, lead concentrations after 6 and 16 hours stagnation are expected to be fairly similar. In consequence of the wide range in this ratio, the shape of the lead dissolution curve can differ markedly and appears to be specific to individual waters. The predicted data in Tables 6.2 and 6.3 are based on lead dissolution curves that have 6 hour–30 minutes stagnation ratios of 2.4 to 3 and are therefore illustrative of the more average condition. Further work will be required to reconcile the model with the Utility's survey results.

(c) Risk assessment

On the basis that the model was adequately calibrated, it was possible to predict the results of random daytime sampling for the water supply system and to predict the percentage of samples that would exceed the WHO guideline value of 10 μg/l. The percentage of failing RDT samples equates to the percentage of population at risk on this basis. The simulations assumed that 14% houses had a lead service pipe. For the present situation and the reduced plumbosolvency scenario, the average percentage of simulated RDT samples exceeding 10 μg/l was determined from 1000 surveys each with 100 samples. The results are shown in Table 6.4.

Table 6.4 Predicted RDT survey results*.

Plumbosolvency		Predicted average RDT samples
M	*E*	% >10 μg/l
0.026	31	0.81
0.010	15	0.02

*Predicted average from 1000 surveys, each of 100 samples.
M = Initial mass transfer rate (μg/m^2/s); *E* = equilibrium concentration (μg/l).

Again, it is interesting to compare these predicted results to what has been achieved in the UK by optimised orthophosphate dosing. At anticipated values of *M* and *E* in the region (i.e. both above and below) of 0.02 and 30, respectively, non-compliance across England and Wales is now down to 1% and in several regions it is down to 0.5% (Hayes & Hydes, 2012). This level of optimisation is being achieved in City B by pH elevation alone. In City B, the percentage of homes with a lead service line (14%) is lower than the average percentage in England and Wales (40%) and is a further factor in determining risk.

6.4 DISCUSSION

It is interesting to compare City A to City B. In both cities water supplies are pH elevated but not orthophosphate dosed and corrosion scales are dominated by Pb(II) compounds. The pH in the City A distribution system averages around 9.8, slightly higher than in City B. The total organic carbon in City A averages around 2 mg/l, slightly lower than in City B. The alkalinity in City A is a little lower at around 20 mg/l (as $CaCO_3$). The free chlorine residual in City A is typically 0.4 mg/l, whereas chlorine is present in the monochloramine form in City B. It is therefore intriguing that the plumbosolvency of the water in City B appears to be significantly lower than in City A, as illustrated by Table 6.5, on the basis of simulated 30MS sampling and the values used for *M* and *E*.

Table 6.5 Comparing City B to City A.

	Average % 30 MS samples >10 μg/l			
	1st Litre	2nd Litre	3rd Litre	4th Litre
Predicted for City B with *M* = 0.026 & *E* = 31	0.21	1.92	7.79	15.23
Predicted for City A with *M* = 0.047 & *E* = 70	4.17	34.39	46.36	28.50

If the values of *M* (0.026) and *E* (31) for City B are correct (or close), then the assumptions about pipe lengths and diameters for lead and non-lead pipes are also likely to be close to actuality. On this basis the case for supplementary orthophosphate dosing would be weak.

However, if the values of *M* and *E* for City B were found to be higher, as might be determined by laboratory plumbosolvency testing, it would cast into doubt the assumptions about pipe lengths and diameters, which are key drivers in determining compliance. Shorter lead pipe lengths or longer non-lead pipes would be implied if predicted results were to be matched to those observed. Another variable that might be implicated would be the percentage of houses that had a lead service line. If significantly less than 95% of houses sampled had a lead service line in the surveys undertaken, it would reduce non-compliance.

It can be noted here that the non-lead pipe lengths and diameters are more relevant to the profile of sequential sample results but do not determine the amount of lead dissolving into the water. If *M* and *E* were found to be higher, lead concentrations at houses with lead pipes would also be higher, with attendant greater risk to the health of infants. In this case, the case for supplementary orthophosphate dosing would be stronger.

The Utility in City B recognises that orthophosphate dosing could reduce plumbosolvency further, but the case for doing this is weak, recognising the high levels of compliance with Provincial standards and Health Canada's guidelines based on 30MS sampling. The Utility also has concerns about biofilm fouling and the destabilisation of corrosion scales although these concerns are probably unfounded on the basis of widespread UK orthophosphate dosing experience (IWA, 2010).

6.5 CONCLUSIONS

(1) Computational modelling has been demonstrated to offer further insights into plumbosolvency and its control in City B.
(2) Laboratory testing to confirm the current plumbosolvency of City B's water supplies and orthophosphate dose responses at different pH values could be considered.
(3) The Utility should consider obtaining basic details of service lines and premise plumbing when the opportunities arise.
(4) The study has revealed apparent inconsistencies in Health Canada's guidelines in the use of 6+ hrs stagnation sampling compared to 30MS sampling. Further elaboration of these inconsistencies is warranted.

Chapter 7

Case study: City C (US)

7.1 BACKGROUND

The emphasis of this case study was the sequential sampling of houses after 6+ hours water stagnation to investigate the behavioural characteristics of the sampling protocol used by the US Lead Copper Rule (LCR). It is also relevant to Tier 1 & 2 assessments in Health Canada's guidelines. The sequential sampling surveys were undertaken in 2011 by the US Environmental Protection Agency to investigate LCR compliance issues in City C. The water in City C is surface derived and supplies have routinely complied with the current regulatory requirements of the LCR which are based on the first litre drawn after stagnation. The water supplies have a moderate alkalinity of about 100 mg/l (as $CaCO_3$) and a pH in the range 7.5 to 7.8. Organic quality appears fairly good on the basis of the reported TOC of 1.5 mg/l. Drinking water of this quality would be expected to have a moderate plumbosolvency with values of M in the range 0.05 to 0.1 and values of *E* in the range 75 to 150. Despite free chlorine residuals in the City of up to 1 mg/l, corrosion deposit analysis by the EPA revealed Pb(II) compounds to be dominant; these have a higher solubility than Pb(IV) compounds.

Until recently, a proprietary ortho-polyphosphate blend has been dosed at about 1 mg/l (as PO_4) resulting in typical orthophosphate concentrations in the City's distribution network of about 0.5 mg/l (as PO_4). Orthophosphate is now being dosed to the City's water supplies at the two treatment plants. It should be noted that polyphosphate is a metal chelating agent and may actually exacerbate plumbosolvency (IWA, 2010). Also, from widespread UK experience, the concentration of orthophosphate required for optimal reduction in plumbosolvency could be about 3 mg/l (as PO_4) for the type of water quality in City C.

7.2 ASSESSMENT OF LEAD DATA FROM SEQUENTIAL SAMPLING SURVEYS

(a) Results profiles

Sequential sampling survey data for 29 sites were provided from two surveys. After flushing for 5 minutes and at least 6 hours' stagnation, 12 or more one litre samples were taken in sequence from each site. One survey was undertaken in June 2011 (28 sites) and the other in September to October 2011 (29 sites). Pipe-work details were available for the sites, as follows: (i) LSL lengths, all; (ii) premise pipe materials, 18; (iii) premise pipe lengths, 21; (iv) giving in combination, information for all of (i) to (iii) for 18 sites.

Over the two surveys, the lead concentrations that were observed varied from 2 to 37 μg/l. All first litre samples had a lead concentration below the Action Level of 15 μg/l. However, in the June 2011 survey, 13 out of the 28 sites exceeded 15 μg/l in at least one sequential sample and in the September to October 2011 survey, 16 out of the 29 sites exceeded 15 μg/l in at least one sequential sample. Figure 7.1 illustrates some of the trends in sequential sample results that were obtained.

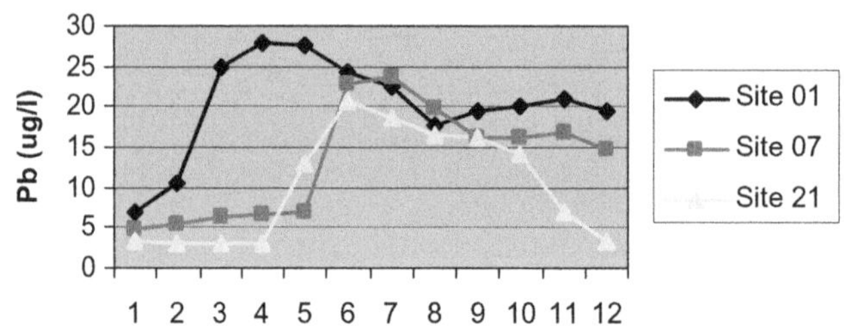

Figure 7.1 Examples of sequential sampling results in City C.

At some sites, elevated lead concentrations were observed for greater parts of the 12-sample sequence than can be explained by the lead service line (LSL) volumes, if plug flow is assumed. If ¾ inch diameter is assumed for the lead service lines, then the following examples illustrate a marked skewing effect:

Site 1: Elevated lead (>15 μg/l) in L3 (i.e. 3rd litre) to L12, a total sample volume of 10 L, compared to the LSL volume of 7.0 L.
Site 8: Elevated lead (>15 μg/l) in L3 to L12, a total sample volume of 10 L, compared to the LSL volume of 4.5 L.
Site 12: Elevated lead (>15 μg/l) in L5 to L11, a total sample volume of 7 L, compared to the LSL volume of 4.1 L.

From the above three examples, it appears that this skew effect can be as high as a factor of 2, implying laminar flow influences during sampling, and that concentrations of lead under plug flow (which is similar to the turbulent flow that could be expected during normal use of the faucet) could therefore be up to twice as high. Sequential sampling is investigated further in Chapter 8.

(b) LCR compliance

LCR surveys from parts of 1992–2008, 14 in total, exhibited an average 90th percentile concentration of 9.2 μg/l (ranging 4 to 20 μg/l), the single non-compliance being in 1992.

From the sequential sampling during 2011, a compliance profile, based on 28 sites sampled in June and 29 sites in September/October, was generated for the first four sequential samples (Table 7.1); the first four samples are of greatest interest as they include Tier 2 (L2–L4) in Health Canada's (2009) guidance that has relevance to the optimisation of plumbosolvency control.

Table 7.1 90th percentile Pb concentrations (*) from Litre 1 to Litre 4, calculated from sequential sampling survey results.

Survey	90th% Pb concn. (μg/l)			
	L1	L2	L3	L4
June 2011	8.30	10.50	18.80	17.70
Sept/Oct 2011	9.19	12.80	16.40	18.90
Average	8.75	11.65	17.60	18.30

*4th highest result in each survey.

This synthesised survey data indicates LCR compliance based on the first litre sample and non-compliance from the third litre on-wards, consistent with the trend shown in Figure 7.1 for site 01. The failure profile based on 90th percentile concentrations can be compared (Table 7.2) to the two other case studies and their plumbosolvency characteristics, as observed or predicted and quantified by M (the initial mass transfer rate in $\mu g/m^2/s$) and E (the equilibrium concentration in μg/l).

The results for each City have been strongly influenced by the lengths of non-lead pipe-work (premise plumbing) that have been used in modelling. None-the-less, the failure profile from City C appears to be associated with a plumbosolvency condition lower than City B (which is itself fairly low). This is in contradiction with the water quality characteristics of the water supplies in City C, which appear likely to be more plumbosolvent, but could be possibly explained by a greater magnitude in the dilution effect from premise plumbing. Inspection of the data on premise plumbing for 21 sites, that were part of the sequential sampling surveys, revealed (Table 7.3) that over 80% had premise plumbing volumes greater than 2 litres and that the average volume was 3.6 litres, thereby exerting a potentially significant distortion effect.

Table 7.2 Comparing City C to other Cities.

	Average 90th percentile Pb concentrations after flushing and 6 hrs stagnation (μg/l)			
	L1	**L2**	**L3**	**L4**
City C observed from sequential sampling (*M* & *E* unknown)	8.75	11.65	17.60	18.30
City A predicted average ($M = 0.047$, $E = 70$)	20.31	64.25	67.42	59.93
City A observed	20.1	N/A	N/A	N/A
City B predicted average ($M = 0.026$, $E = 31$)	0.09	20.20	30.54	30.62

N/A = not available.

Table 7.3 Premise plumbing volumes.

Volume of premise plumbing (litres)	Number of sites	Percentage of sites
<1	2	9
1 to 2	2	10
2 to 3	4	19
3 to 4	3	14
>4	10	48

If applicable to City C more widely, then the influence of dilution from non-lead premise plumbing extends significantly further than the first litre sample. It is therefore possible that the plumbosolvency of City C's water supplies is significantly higher than is apparent from the synthetic compliance data summarised in Table 7.1.

7.3 LEAD SERVICE LINES AND PLUMBOSOLVENCY CHARACTERISATION

According to the EPA, all service connections of 2 inch diameter or smaller were made of lead until 1986 and most residential lead service lines in City C are either ¾ or 1 inch in diameter. About 85% of residential properties are likely to have been built on or before 1986, in line with general trends in housing. This means that about 85% of residential properties in City C could have a lead service line (LSL), notwithstanding any partial LSL replacement. Dependent on the plumbosolvency of the water supplies in City C, up to about 85% of homes could therefore be at risk from lead in drinking water. It is possible that elevated concentrations of lead in drinking water are widespread in City C and have been greatly under-estimated by the LCR sampling protocol. Laboratory testing should be considered to determine the level of plumbosolvency of the City's supplies, using the method that was deployed by the great majority of UK water companies (IWA, 2010); results can be obtained within a month.

7.4 MODELLING

(a) Zonal compliance modelling

Presently, too little data has been provided to be able to calibrate the zonal compliance model to the extent possible for Cities A and B. However, a speculative exercise has been undertaken, based on the pipe-work characteristics of Cities A and B and assuming plumbosolvency factors ranging from $M = 0.03$ to 0.05 and $E = 45$ to 75. These plumbosolvency factors reflect: (i) the lead concentrations observed in the sequential sampling surveys, and (ii) reference to extensive UK data from laboratory testing for low orthophosphate doses to moderately high alkalinity waters.

The simulation results for 6 hr stagnation sampling are shown in Table 7.4. Neither City A nor B results map on to the synthesised data for City C. It is possible that this is because the synthesised data for City C has been influenced to a greater extent by the length of non-lead premise plumbing (with reference to Table 7.3). The results derived from City A are generally higher than from City B due to the generally shorter non-lead pipe lengths that were assumed in City A, exerting less dilution.

Table 7.4 Simulated 6 hr stagnation surveys for City C based on City A and B characteristics for a range of plumbosolvency conditions*.

	Average 90th percentile concentrations (μg/l)			
	L1	**L2**	**L3**	**L4**
City C synthesised from sequential sampling	8.75	11.65	17.60	18.30
Predicted for City C based on City A ($M=0.05$, $E=75$)	16.60	67.60	71.04	58.78
Predicted for City C based on City A ($M=0.04$, $E=60$)	13.94	52.52	56.75	44.89
Predicted for City C based on City A ($M=0.03$, $E=45$)	12.31	40.18	42.11	34.02
Predicted for City C based on City B ($M=0.05$, $E=75$)	0.19	46.39	72.73	72.88
Predicted for City C based on City B ($M=0.04$, $E=60$)	0.13	38.14	58.13	58.26
Predicted for City C based on City B ($M=0.03$, $E=45$)	0.10	28.37	43.59	43.70

*Average of simulated 1000 surveys, each of 100 samples.

The simulation results for random daytime (RDT) sampling are shown in Table 7.5 and clearly indicate the relationship between plumbosolvency (as quantified by M and E) and the risk of exceeding the WHO Guideline of 10 μg/l. These results assume that 85% of houses in City C have a lead service line, which is subject to confirmation.

Table 7.5 Predicted RDT survey results for City C*.

M	*E*	% >10 μg/l Based on City A	% >10 μg/l Based on City B
0.05	75	12.46	18.60
0.04	60	8.23	11.94
0.03	45	4.34	7.67

*Average of 1000 simulated surveys, each of 100 samples.

To utilise these computational modelling tools further will require better calibration data. Better indicative results could quickly be obtained using assumptions about pipe-work circumstances if the level of plumbosolvency could be determined by laboratory testing. It would also be necessary to firm up on an estimate of the percentage occurrence of lead service lines in the City. This would enable a better assessment of the extent of plumbosolvency problems in City C to be determined, together with a better understanding of the effectiveness of orthophosphate dosing.

(b) Additional modelling to investigate laminar flow effects

Section 3 provided the initial results of CFD modelling. These initial results indicated skewing effects, associated with laminar flow, of the same magnitude as those observed at several sites sampled in all three Cities. Chapter 8 further investigates the behavioural characteristics of sequential sampling.

7.5 DISCUSSION

Much of the assessment of the lead in drinking water situation in City C has focused on two sequential sampling surveys undertaken in 2011. The observational data not only has relevance to LCR compliance issues (which are actively under review by the EPA), but can be used more broadly, to some extent, in the characterisation of the lead in drinking water situation in City C.

The plumbosolvency of water supplies in City C is presently vague but the circumstantial evidence available suggests that there might be scope to reduce lead concentrations by increasing the dose of orthophosphate. Laboratory plumbosolvency testing could quickly clarify the situation and provide a firmer basis for compliance modelling. The information provided on the use of lead pipes by the Utility up to 1986 suggests that a large proportion of the City's housing stock, perhaps as much as 85%, has a lead service line. Dependent on the plumbosolvency of the water supplies in the City, this could represent a significant risk to consumers, particularly children. This position needs to be clarified.

7.6 CONCLUSIONS

(1) Only limited data was available to calibrate the compliance model, but speculative simulations have demonstrated how modelling can provide a deeper insight into plumbosolvency control.
(2) The extent of occurrence of lead service lines in the City should be clarified and information sought on their lengths and diameters.
(3) More information on premise plumbing should also be sought, particularly pipe materials, lengths and diameters.
(4) Laboratory testing could quickly clarify the plumbosolvency characteristics of the City's water supplies.
(5) From (2) to (4) above, the extent of lead in drinking problems in the City could then be clarified.

Chapter 8

Investigations into sequential sampling

8.1 INTRODUCTION

Presently, there is no direct regulatory significance of sequential sampling in the US, although this might change as the US Lead Copper Rule is actively under review. In Canada, sequential sampling after both 30 minutes and 6+ hours stagnation is used as a basis for the optimisation of plumbosolvency control (Health Canada, 2009). In at least one Canadian Province, sequential sampling of the first two litres after 30 minutes stagnation is the basis for regulatory compliance.

As identified in previous sections, the skewing of lead concentrations in the sequence of samples taken after stagnation is significant and most likely variable, making it difficult for the prescriptive use of sequential sampling to be included in regulatory instruments. It could also limit the use of sequential sampling within optimisation protocols.

This section provides further information on the characteristics of sequential sampling by utilities and related agencies, and the results of an initial modelling investigation.

8.2 SEQUENTIAL SAMPLING SURVEYS IN CITIES A, B AND C

(a) City A–results of sequential sampling by the State Health Authority

The results of sequential sampling after 6+ hours stagnation in relation to pipe-work characteristics are shown in Table 8.1.

Table 8.1 City A–Sequential sampling prior to partial lead service line replacement*.

Site No.	Pb service volume (litres)	Premise pipe volume (litres)	Total pipe volume (litres)	Premise pipe materials (mainly)	Sequential litre samples with elevated Pb >15 µg/l
1	1.47	2.05	3.52	Brass, some Cu	5 to 7
2	1.66	2.47	3.13	Brass	4 to 7
3	1.20	2.07	3.27	Brass	1 to 6
4	1.50	2.30	3.80	Brass	4 to 6
5	2.40	1.50	3.90	Brass	3 to 7
6	1.30	2.10	3.40	Copper	2 to 4
7	1.00	1.90	2.90	Brass	5 to 6
8	2.60	2.69	5.29	Brass, some Cu	3 to 8

*Pipe volumes were calculated by the State Health Authority based on pipe-work inspections.

These lead results are not consistent with simple plug flow, which is an adequate approximation of the turbulent flow expected to occur in small bore domestic pipes at normal service flow rates. Assuming that the lead service lines were the principal source of lead in the samples, the difference between sample volume with lead >15 μg/l and LSL volume was up to 5 times. If premise plumbing, much of which was made of brass pipes, is included then the difference is up to about 2 times.

(b) City B – results of sequential sampling by the utility

A small selection of sequential sampling results, from City B, is shown in Figure 8.1. Samples were taken after flushing and 30 minutes stagnation. Compliance with the standard of 10 μg/l is evident for the first two litres at all five sites, but lead concentrations then increased above this standard at four sites.

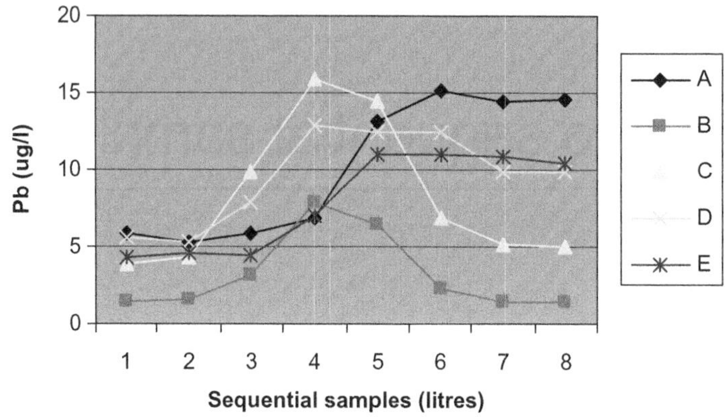

Figure 8.1 Sequential sampling results from City B.

(c) City C – results of sequential sampling by the USEPA

Similar results to those shown for City B in Figure 8.1 have been given for City C in Figure 7.1 in Chapter 7 and show a similar pattern, albeit after 6+ hours stagnation and in relation to a standard of 15 μg/l.

Table 8.2 City C – Sequential sampling in relation to pipe-work*.

Site No.	Pb service volume (litres)	Premise pipe volume (litres)	Total pipe volume (litres)	Premise pipe materials	Sequential litre samples with elevated Pb >15 μg/l
1	7.0	0.6	7.6	Copper	3 to 12 = 10 litres
5	6.3	2.5	8.8	Iron	4 and 5 = 2 litres
6	4.7	5.6	10.3	Iron	0 litres
7	7.4	2.2	9.6	Copper	6 to 11 = 6 litres
8	4.5	4.2	8.7	Iron/Copper	3 to 12 = 10 litres
12	4.1	1.4	5.5	Iron/Copper	5 to 11 = 7 litres
18	6.0	3.8	9.8	Iron	0 litres
19	4.9	2.5	7.4	Copper/Iron	0 litres
22	5.1	6.3	11.4	Iron	0 litres
23	5.2	3.3	8.5	Iron	0 litres
25	5.5	4.9	10.4	Iron	0 litres
26	5.2	4.5	9.7	Iron	0 litres
29	12.4	0.3	12.7	Copper	2 to 8 = 7 litres
31	5.6	4.1	9.7	Copper	6 to 10 = 5 litres
32	3.4	5.5	8.9	Iron	6 = 1 litres
35	4.0	2.2	6.2	Iron/Copper	0 litres

*Data for 16 out of 28 sites for which the data-set was available relating to the survey in June 2011; it has been assumed that all pipes were 18 mm diameter.

The results of sequential sampling after 6+ hours stagnation in relation to pipe-work characteristics are shown in Table 8.2 for 16 sites from the survey undertaken in June 2011. All first litre samples were below the Action Level of 15 µg/l. The relationship between elevated lead concentrations (>15 µg/l) and the length of lead service lines was not clear. In a few cases, what might be described as "background" lead was evident at around 3 to 4 µg/l, suggesting the absence of a lead service line. In other cases, lead concentrations were elevated in some parts of the sample sequence, but remained below 15 µg/l. An alternative explanation is that: (i) the ortho-polyphosphate blend was effective at some sites; and (ii) the higher background lead concentrations at other sites were due to effects from past disturbance of the pipework. Further more detailed investigations will be needed if the variable results from the sequential sampling surveys are to be better understood.

8.3 MODELLING ZONAL COMPLIANCE

It is clear that the results for lead from sequential sampling have a tendency to be skewed, albeit in a variable manner. Skewing can also be demonstrated by the zonal compliance model which can simulate plug flow (equivalent to turbulent flow) or laminar flow, as selected. The results in Figure 8.2 are the average simulated 90th percentile concentrations for each sequential sample (i.e. first litre, second litre, etc.) after 30 minutes stagnation for a zone with low plumbosolvency water ($M = 0.021$, $E = 26$, based on City B characteristics). Under simulated laminar flow, the average 90th percentile lead concentrations almost halved compared to plug flow and tapered off more slowly. These trends have much significance to the manner in which sequential sampling survey results are interpreted in the optimisation of plumbosolvency control treatment.

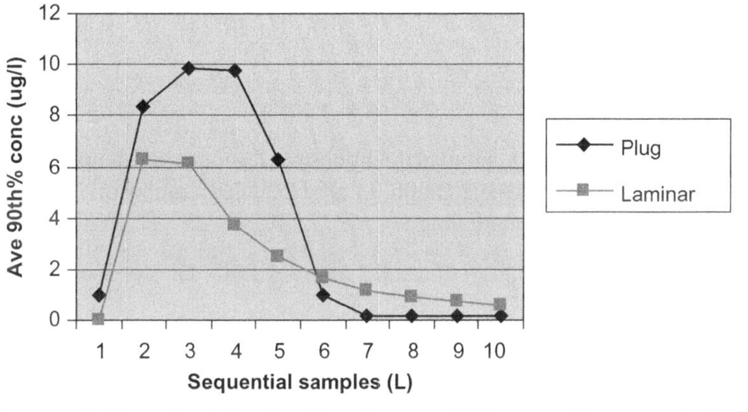

Figure 8.2 Skewing of lead concentrations due to laminar flow. Average 90th percentile concentrations based on 1000 simulated surveys, each of 100 samples; $M = 0.021$ and $E = 26$.

8.4 USING REYNOLD'S NUMBER

Whether or not flow is likely to be turbulent, transitional or laminar can be estimated from Reynold's Number (Re), where:

$$\mathrm{Re} = \frac{\mathrm{Flow}\,(\mathrm{m}^3/\mathrm{s}) \times \mathrm{Diamter}\,(\mathrm{m}) \times 10^6}{\mathrm{Area}\,(\mathrm{m}^2)}$$

Flow is considered turbulent if $\mathrm{Re} \geq 4000$ and laminar if $\mathrm{Re} \leq 2000$. Table 8.3 indicates the likely flow type for 12 and 18 mm diameter pipes for a range of flows.

Table 8.3 Reynold's numbers.

Diameter (mm)	Flow (litres/sec)	Flow (UK gall/min)	Flow (US gall/min)	Re	Flow type
12	0.10	1.32	1.58	10,610	Turbulent
12	0.02	0.26	0.22	2,120	Transitional
18	0.10	1.32	1.58	7,070	Turbulent
18	0.02	0.26	0.22	1,415	Laminar

It is possible that the normal household use of the faucet is associated with higher flows than during sequential sampling. Further research will be needed to clarify this position (not part of this study). The compliance model does not automatically select the plug or laminar flow modelling component, but must be selected by the user. In this study, the plug flow model was used for compliance assessment in the case studies, in line with the numerous earlier cases in the UK. This most likely remains the correct approach when simulating the first litre drawn from the faucet but for sequential sampling, both the plug and laminar flow models could be used to cover the range of conditions that are possible.

8.5 MODELLING SEQUENTIAL SAMPLING AT A SINGLE HOUSE

(a) Introduction

The single pipe model was used to investigate the lead emission characteristics of different pipework circumstances at a single house, comprising a simulated lead pipe (service line) and a simulated copper pipe (premise plumbing) of various lengths and diameters, for both plug and laminar flow conditions. The number of observed cases from the sequential sampling surveys undertaken, that could be used to validate the model, was limited because:

- all but one site surveyed in City A had brass piping within the premises and there was evidence of lead contributions from premise plumbing in City A, possibly due to brass components; and
- in City C, lead emissions were variable and did not correlate generally with the pipework details that were available.

Despite uncertainties about lead contributions from brass pipes, two cases from City A where brass pipes were present were included in the validation exercise.

(b) Validation exercise

The three case examples from City A, in which simulated and observed sequential sample results have been compared, are shown in Figures 8.3 to 8.5, based on 6+ hours stagnation. In all cases, the plumbosolvency factors $M = 0.047$ and $E = 70$ were applied (as used in the LCR compliance modelling).

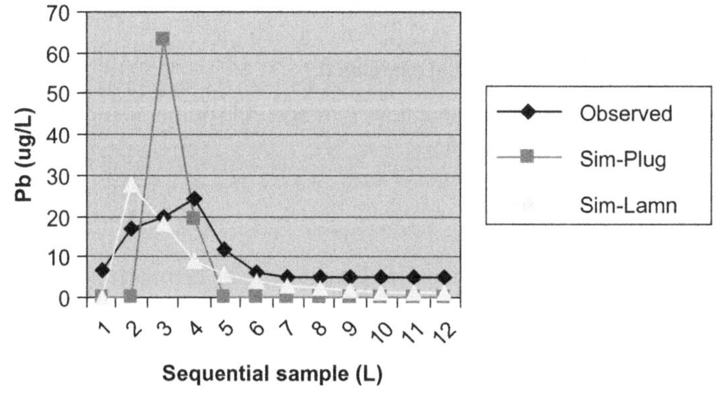

Figure 8.3 Predicted and observed sequential sample results at Site 6 with lead and copper pipes (State Health Authority survey). *Note*: Lead pipe: 9.0 m length, 13 mm diameter; copper pipe: 15.7 m length, 13 mm diameter.

It can be seen from Figures 8.3 to 8.5 that the predicted results for laminar flow compare more closely to the observed results than for plug flow. For Site 6 and Site 2 both the predicted and observed results relate to the faucet. The match between predicted and observed results was closer for Site 6 than Site 2 under laminar flow conditions. For Site 2, the match between predicted and observed results was closer under plug flow conditions. For Site 109PR, the sample sequence was taken from the outlet position of the brass pipe, not the faucet, due to a further section of lead pipe (4.0 m length) and a section of copper pipe (1.3 m length) being present adjacent to the faucet that were beyond the scope of the model. The validation that has been achieved was considered sufficient to enable the model to be used for investigative purposes and the results of three investigations are summarised below.

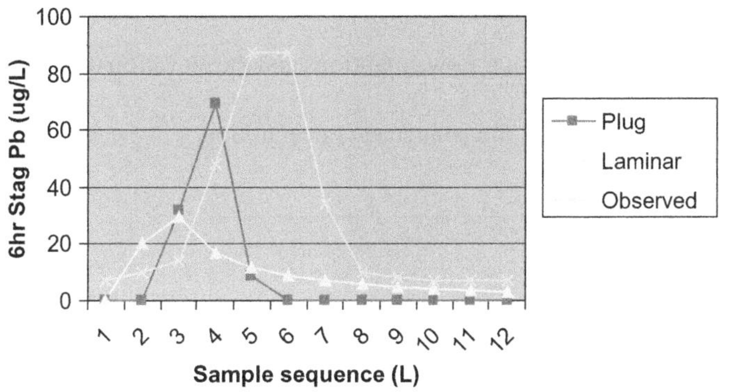

Figure 8.4 Predicted and observed sequential sample results at Site 2 with lead and brass pipes (State Health Authority survey). *Note*: Lead pipe: 11.9 m length, 13 mm diameter; brass pipe: 16.5 m length, 14 mm diameter.

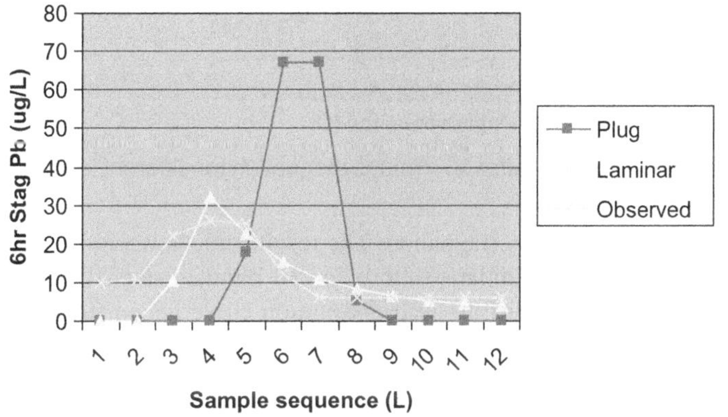

Figure 8.5 Predicted and observed sequential sample results at Site 109PR with lead and brass pipes (Utility survey). *Note*: Lead pipe: 9.2 m length, 18 mm diameter; brass pipe: 18.6 m length, 18 mm diameter.

(c) Effect of copper pipe length

The length and diameter of the lead pipe were held constant at 15 m and 18 mm, respectively. The length of the copper pipe was varied although the diameter was held constant at 18 mm. In all cases the plumbosolvency factors were set at $M = 0.05$ and $E = 75$. The results for 30 minutes stagnation and plug flow are shown in Figure 8.6 from which it can be determined that:

- the length of the copper pipe markedly affected the lead emission profile of the sequence of twelve one-litre samples
- the longer the copper pipe, the later was the sequence of elevated lead concentrations associated with the lead pipe

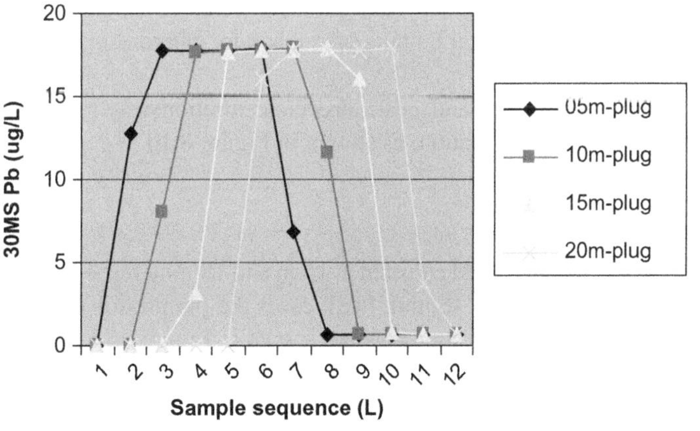

Figure 8.6 Predicted sequential sample results after 30 minutes stagnation for different lengths of copper pipe with plug flow

The results for 30 minutes stagnation and laminar flow are shown in Figure 8.7 from which it can be determined that:

- lead concentrations were lower under laminar flow conditions and the elevation of lead concentrations was spread over a greater number of sequential samples
- the longer the copper pipe, the lower was the peak lead concentration

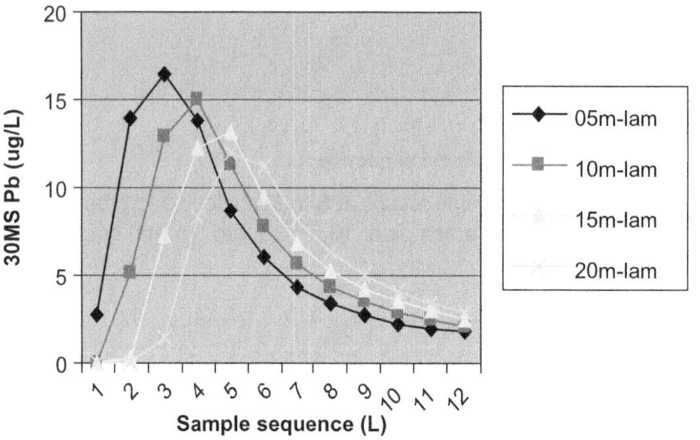

Figure 8.7 Predicted sequential sample results after 30 minutes stagnation for different lengths of copper pipe with laminar flow.

Simulations were also undertaken for 6 hours stagnation. The results are shown in Figures 8.8 and 8.9 for plug flow and laminar flow, respectively. The results were similar to those for 30 minutes stagnation except that the concentrations of lead were about four times higher, a reflection of the lead dissolution curve generated by M and E. The conclusions for 6 hours stagnation are therefore the same as for 30MS.

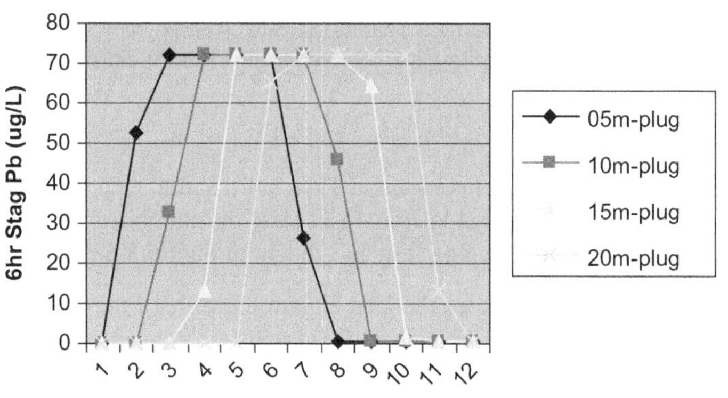

Figure 8.8 Predicted sequential sample results after 6 hours stagnation for different lengths of copper pipe with plug flow.

The relationship between copper pipe length and peak lead concentration was found to be linear under laminar flow conditions, for both 30 minutes and 6 hours stagnation, as shown in Figure 8.10.

(d) Effect of lead pipe length

The length and diameter of the copper pipe were held constant at 15 m and 18 mm, respectively. The length of the lead pipe was varied although the diameter was held constant at 18 mm. In all cases the plumbosolvency factors were set at $M = 0.05$ and $E = 75$. The results for 30 minutes stagnation and plug flow are shown in Figure 8.11 from which it can be determined that:

- the length of the lead pipe markedly affected the lead emission profile of the sequence of twelve one-litre samples
- the longer the lead pipe, the more extensive was the sequence of elevated lead concentrations associated with the lead pipe

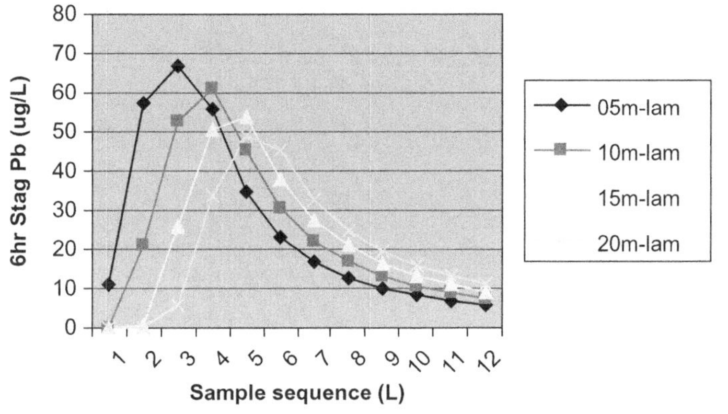

Figure 8.9 Predicted sequential sample results after 6 hours stagnation for different lengths of copper pipe with laminar flow.

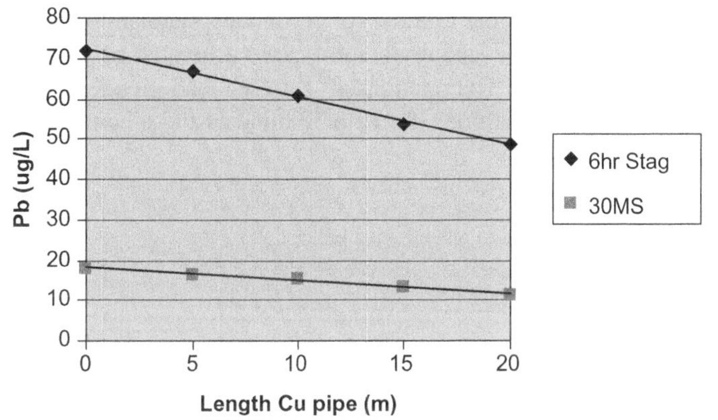

Figure 8.10 Relationship between copper pipe length and peak lead concentration under laminar flow conditions.

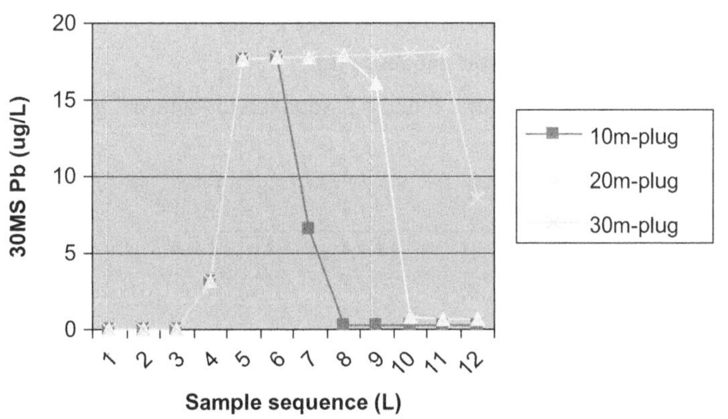

Figure 8.11 Predicted sequential sample results after 30 minutes stagnation for different lengths of lead pipe with plug flow.

The results for 30 minutes stagnation and laminar flow are shown in Figure 8.12 from which it can be determined that:

- lead concentrations were lower under laminar flow conditions and the elevation of lead concentrations was spread over a greater number of sequential samples
- the longer the lead pipe, the higher was the peak lead concentration

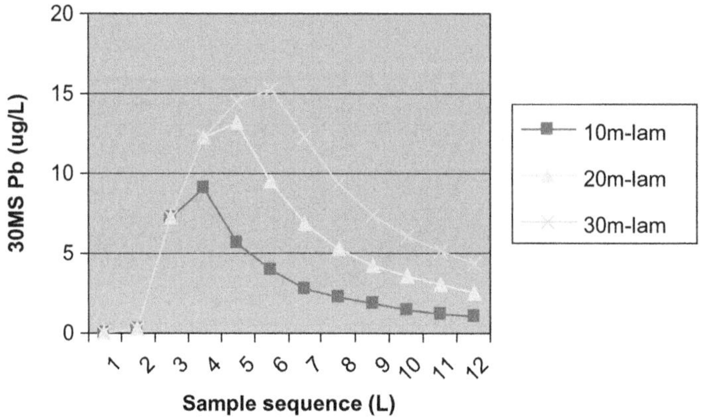

Figure 8.12 Predicted sequential sample results after 30 minutes stagnation for different lengths of lead pipe with laminar flow.

Simulations were also undertaken for 6 hours stagnation. The results are shown in Figures 8.13 and 8.14 for plug flow and laminar flow, respectively. The results were similar to those for 30 minutes stagnation except that the concentrations of lead were about four times higher, a reflection of the lead dissolution curve generated by *M* and *E*. The conclusions for 6 hours stagnation are therefore the same as for 30MS. The relationship between lead pipe length and peak lead concentration was found to be curvi-linear under laminar flow conditions, for both 30 minutes and 6 hours stagnation, as shown in Figure 8.15.

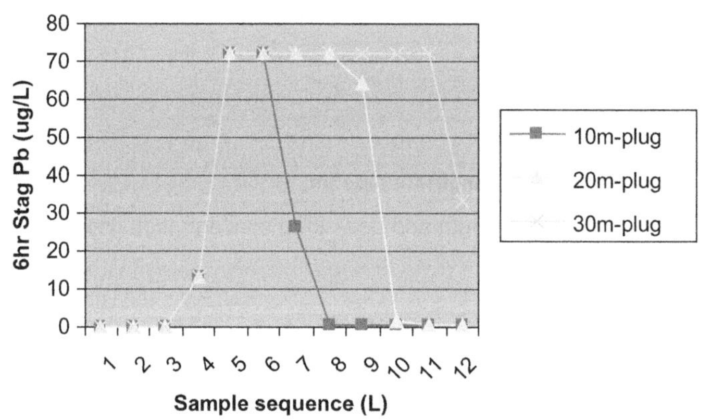

Figure 8.13 Predicted sequential sample results after 6 hours stagnation for different lengths of lead pipe with plug flow.

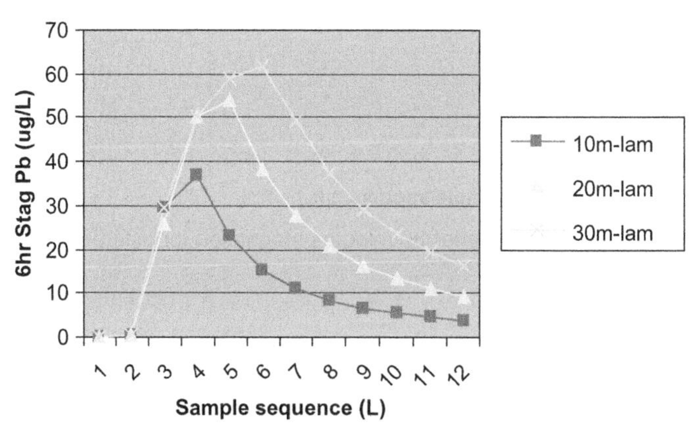

Figure 8.14 Predicted sequential sample results after 6 hours stagnation for different lengths of lead pipe with laminar flow.

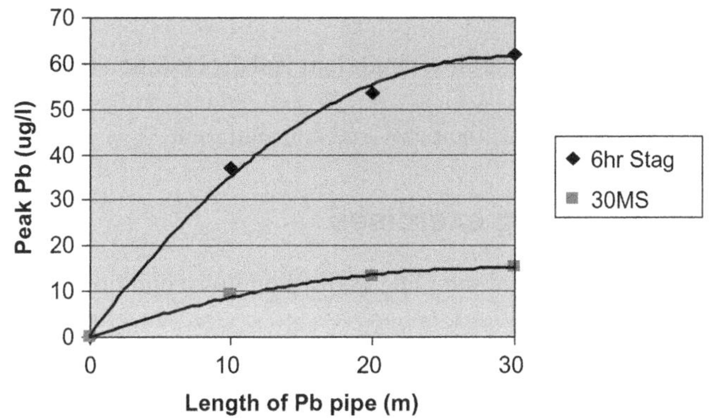

Figure 8.15 Relationship between lead pipe length and peak lead concentration under laminar flow conditions.

(e) Effect of pipe diameters

The lengths of both the lead pipe and the copper pipe were held constant at 15 m. The diameters of the pipe were then varied. In all cases the plumbosolvency factors were set at $M = 0.05$ and $E = 75$.

The results for 30 minutes stagnation and plug flow are shown in Figure 8.16 from which it can be determined that:

- the diameters of the pipes markedly affected the lead emission profile of the sequence of twelve one-litre samples
- the larger the diameters, the more extensive was the sequence of elevated lead concentrations associated with the lead pipe
- the larger the diameters, the lower was the concentration of peak lead

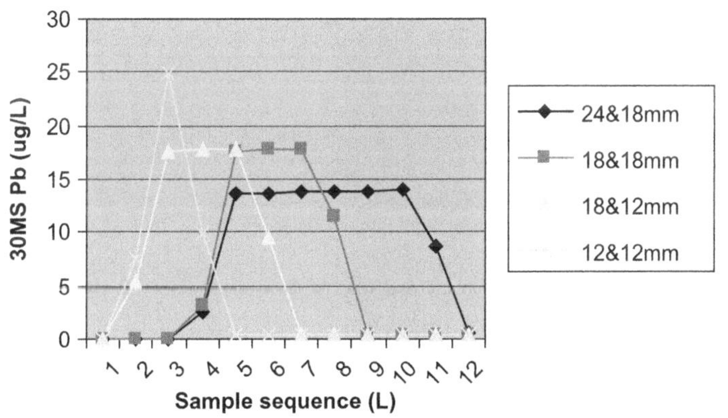

Figure 8.16 Predicted sequential sample results after 30 minutes stagnation for different pipe diameters with plug flow. First diameter shown is key relates to the lead pipe and the second to the non-lead pipe.

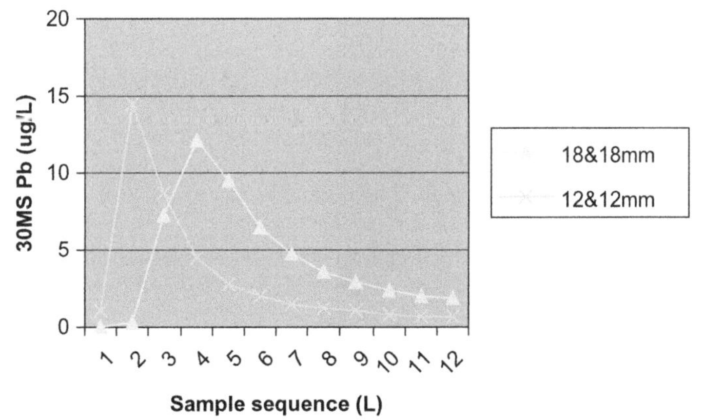

Figure 8.17 Predicted sequential sample results after 30 minutes stagnation for different pipe diameters with laminar flow. First diameter shown in key relates to the lead pipe and the second to the non-lead pipe.

The results for 30 minutes stagnation and laminar flow are shown in Figure 8.17 from which it can be determined that:

- lead concentrations were lower under laminar flow conditions and the elevation of lead concentrations was spread over a greater number of sequential samples
- the larger the pipe diameters, the lower was the peak lead concentration

(f) Conclusions from the modelling exercises

(1) The lead emission profiles from sequential sampling under plug flow conditions are markedly different to those under laminar flow conditions.

(2) It appears from the monitoring exercises undertaken in Cities A, B and C that laminar flow conditions have had a stronger influence on the results obtained.

(3) The results from the single pipe model under laminar flow conditions matched fairly well the results from actual surveys at two sites. At a third site, the match was better under plug flow conditions. It should be borne in mind that the flow conditions at an individual property will be somewhere in between the idealised plug and laminar conditions that have been reported.

(4) Modelling revealed that the sequence of sample results and peak lead concentrations were markedly affected by the length of lead pipe, by the length of copper pipe and by pipe diameters.

(5) In consequence, sequential sampling is far too subject to variable influences to be used in definitive terms for either regulatory compliance assessment or for the optimisation of plumbosolvency control measures, if sampling is undertaken from different sets of houses from within a changing sampling pool.

(6) However, repeated sequential sampling at specific locations should be able to provide a performance benchmark for the optimisation of plumbosolvency control measures, but only if flow rates during sampling are kept constant. In such cases, the lead concentrations within the profiles must be sufficiently high to be able to demonstrate treatment changes.

(7) Sequential sampling can also be used for diagnostic purposes, but differentiating sources of lead can be affected by skewing and lowering effects.

Chapter 9

Discussion

9.1 THE USE OF MODELLING IN THE OPTIMISATION OF PLUMBOSOLVENCY CONTROL

(a) The limitations of sampling

The fundamental problem in the optimisation of treatment measures for plumbosolvency control is the absence of a simple and rapid control loop. The inherent variability of lead emissions from plumbing systems at their point of use, the faucet, challenges all sampling methods and monitoring protocols. Problems extend to poor reproducibility dependent on sample numbers, distortion from the dilution effects of water stood in non-lead pipe-work, skewing of sample results due to laminar flow influences, the possibility of poor sampling technique when consumer volunteers are involved, seasonal variation (both temperature and water quality components such as organics), changes over time in the sampling pool, and any inconsistencies in the control measures such as fluctuations in pH or the concentration of corrosion inhibitor. The use of lead pipe test rigs or loops is further complicated by the different behavioural characteristics of the old exhumed or in-situ lead pipes compared to new lead piping. Such problems may not be fully appreciated by those involved, particularly as detailed experimental studies on sampling and monitoring are limited to a fairly small number of published case studies and research projects. Part of this limitation is due to the time and resources required in experimental investigations, even when sampling and monitoring features so obviously as a component of regulatory mechanisms.

(b) The use of computational modelling tools

Computational modelling can investigate the behavioural characteristics of sampling and monitoring protocols, and exhaustive investigations can be undertaken quickly at low cost. However, there are problems of perception in the potential user-community that need to be overcome. The answer lies in case studies, such as those in this project, that can be related to by the practitioners involved. In building acceptance, the performance attributes of the model that had to be addressed were:

- The sensitivity of the model to the calibration inputs – in this project, the focus here was on the plumbosolvency factors *M* and *E*;
- Variance in the model's results – this was demonstrated to be limited to a low level;
- Resolution – the results for the different scenarios investigated were readily able to be differentiated; and
- Plausibility – as far as possible the calibration inputs (such as pipe-work lengths) were those provided by the research collaborators or equivalent points of reference, and amendments were kept to a minimum.

This project commenced on a firm basis due to the model's extensive and successful track record from the optimisation of numerous water supply systems in the UK. The calibration data on non-lead pipe-work is not a major influence on the simulation of random daytime samples, as are used in the UK to assess regulatory compliance. However, in the North American context, this calibration data was found to be critical. It is somewhat ironic that this component of the water supply system, that is not the responsibility of the Utility, never-the-less influences the Utility's regulatory position to the extent that it does.

The modelling exercises that investigated the behavioural characteristics of sequential sampling after either 30 minutes or 6 hours stagnation are particularly important and timely. Either sequential sampling has recently been built into national guidelines, or provincial standards, for compliance assessment and for the optimisation of plumbosolvency control treatment (Canada), or is being considered in these terms (US). The modelling exercises indicated that sequential sampling has significant potential limitations and further more detailed research involving case studies, encompassing both modelling and experimental work, could well be warranted.

(c) Supporting techniques

In all three case studies, better quantification of plumbosolvency would have been beneficial. A rapid, low-cost testing technique is well established and is appropriate when Pb(II) compounds dominate the lead pipe corrosion deposits, which is most commonly the case. Plumbosolvency testing linked to compliance modelling will enable likely optimum treatment conditions to be quickly determined, albeit these conditions must then be confirmed by field monitoring. Such testing is rapid because it uses new lead piping whereas old in-situ lead pipes take much longer to equilibrate with any new treatment condition. Much of the other calibration data on pipe-work can be fairly easily acquired if brief inspections were to be linked to routine sampling activities. X-ray diffraction (XRD) analysis of a few lead pipe cut-outs should also be considered.

9.2 REGULATORY ASPECTS

(a) United States

It is very clear from the sequential sampling surveys undertaken in Cities A, B and C, and from the related modelling investigations, that the first litre drawn from the faucet after 6+ hours stagnation is likely to comply with the Action Limit of 15 µg/l, due to dilution effects from water stood in non-lead pipe-work. In consequence, regulatory surveys are likely to comply with the Lead Copper Rule (LCR), that requires the 90th percentile lead concentration to be less than 15 µg/l, thereby under-estimating the extent of lead in drinking water problems that might be being experienced in a water supply system. This situation is relevant to public health protection, particularly infants who are more susceptible to lead poisoning (WHO, 2010). If a water supply system fails to comply with the LCR for lead, then optimised corrosion control is required together with a replacement programme for that part of the lead service line that is owned by the Utility. Such requirements are therefore dependent on the results of a flawed sampling protocol, which cannot be in the best interests of any stake-holder (consumer, Utility or the regulatory agencies).

This flawed situation is made worse because it will vary, as demonstrated by both the results of the LCR surveys undertaken by utilities (ref: City A) and modelling, and it cannot provide any reliable or robust basis for the optimisation of plumbosolvency control. The contribution of lead to drinking water from premise plumbing appeared to be more significant in City A than City C, both of which are in the US, possibly because brass pipes are more common in City A.

The LCR is actively under review and may be revised. One option would be to follow the precedent from Canada of using sequential sampling to help overcome the problems associated with assessing compliance on the basis of the first litre alone. However, this project clearly demonstrates that sequential sampling would not be without problems.

An alternative approach could be to define the meaning of the term "optimisation" (as it applies to plumbosolvency control) in some manner that linked achieving the greatest possible reductions in lead concentrations to operational feasibility and not entailing excessive cost. The modelling tools demonstrated by the project could contribute substantially to this process of optimisation by enabling utilities to quickly explore the relationship between treatment measures and compliance, without incurring the higher costs and longer timescales of field trials (other than to confirm the optimum conditions identified).

(b) Canada

The Health Canada guidelines (2009) for prompting corrective action suffer from the same problems as the US Lead Copper Rule. For residential properties, the 6+ hours stagnation sampling protocol relies initially on the first litre taken after stagnation (Tier 1) from at least 50% of homes with a lead service line. If 15 µg/l is exceeded in the first litre at more than 10% of the sites sampled then the analysis of second, third and fourth litre samples are prompted (Tier 2). As has been demonstrated by both utility surveys and modelling, not only are Tier 1 samples subject to distortion effects but also the Tier 2 samples that follow. Health Canada (2009) indicates that the results from Tier 1 and Tier 2 samples, taken before and after corrosion control measures are implemented, can be used to conduct an assessment of corrosion control efficacy and optimisation. This guidance is put into question by the range of variable effects that have been clearly demonstrated by both utility surveys and by modelling.

The Health Canada guidelines (2009) provide an alternative approach based on 30 minutes stagnation sampling from homes, all of which should have a lead service line. The sampling protocol involves the collection of four sequential litre samples after flushing and stagnation. If the average lead concentration of the four samples taken from more than 10% of homes exceeds 10 μg/l then corrective measures should be implemented. The averaging of the four-sample sets is open to question when distortion can affect all samples taken. Even with regard to the skewing and flattening of results that can occur, it would be more appropriate to use the sample with the highest lead concentration. If amended in this way, routine (weekly or monthly) sequential 30MS sampling at selected properties should be able to provide a benchmarking method that can be used to good effect in the optimisation of plumbosolvency control, subject to the observed lead concentrations being high enough to be capable of demonstrating change, and subject to flow conditions during sampling being constant.

The modelling undertaken in this project has revealed that the compliance and optimisation protocols based on sampling after 6+ hours stagnation are likely to be more stringent than those based on sampling after 30 minutes stagnation. This is not surprising as the numerical difference between the lead concentration after 6 hours stagnation is typically around 3 to 4 time higher than after 30 minutes stagnation, whereas the numerical difference between the standards used is only 1.5 (i.e. 15 μg/l compared to 10 μg/l). However, some of the data, but not all, from City B indicated that lead concentrations after 30 minutes and 6+ hours stagnation were closer. Laboratory plumbosolvency testing has indicated that the shape of the lead dissolution curve can vary (IWA, 2010), which may help to explain these observations.

9.3 OPERATIONAL ASPECTS

A number of operational issues have been raised during the course of the project. Brief comments are offered as follows:

(a) *Transition from high pH to orthophosphate*: there are no published case studies where the pH is in the range 9 to 10, albeit there are plenty up to pH 9 (e.g. IWA, 2010); laboratory plumbosolvency testing could provide an initial indication of the significance of pH on the orthophosphate dose response of the water; pilot trials will be necessary to evaluate whether or not changes to the corrosion deposits within plumbing systems might be problematical.
(b) *Stimulation of bio-films by dosing orthophosphate*: there is no reported evidence of problems occurring in the UK where 95% of supplies are dosed.
(c) *High costs of orthophosphate*: although the costs are high compared to other treatment chemicals, the doses applied are small such that the unit cost is only about 0.5 cents ($) per m^3 for a typical dose of 1 mg/l (P). Water companies in the UK have expressed concern about the security of supply of orthophosphate.
(d) *Environmental impact*: typical doses to water supplies of around 1 mg/l (P) are much lower than the typical 10 mg/l (P) content of sewage; a 10% increase in orthophosphate was considered to be negligible in the UK and no adverse impacts have been demonstrated (IWA, 2010)

9.4 RISK ASSESSMENT

Simulating the results of random daytime (RDT) sampling provides a simple basis for risk assessment using the WHO guideline value of 10 μg/l as a health related benchmark.

City B had the lowest risk level at 0.8% > 10 μg/l and compares closely to the present 1% risk level in England and Wales following the optimisation of orthophosphate dosing. In City B this level of plumbosolvency control is being achieved by pH elevation alone at between 9.2 and 9.4.

City A was found to be intermediate with 5.2% > 10 μg/l and there may be scope for further reductions in lead concentrations, subject to operational feasibility.

City C was found to have the highest risk, between 4.3 and 18.6% > 10 μg/l, depending on the plumbosolvency of the water supplies and the occurrence of lead service lines, both of which were uncertain. The possibility of this higher level of risk should prompt the relevant agencies to investigate more closely.

9.5 THE WAY FORWARD

The main objective of this collaborative research project was to demonstrate the applicability of computational compliance modelling in the North American context following its successful use in the optimisation of plumbosolvency control in the UK. This objective was achieved.

In the optimisation of plumbosolvency control treatment in the three case study Cities, the next step could be to overcome the calibration data limitations that have been identified; much of the uncertainties could quickly be reduced by plumbosolvency testing and by obtaining better information on pipe-work circumstances, particularly premise plumbing. Remodelling could then identify optimum treatment conditions with greater confidence.

Pilot trials could then follow, using appropriate monitoring techniques, to demonstrate the extent of reductions in lead concentrations. In addition to satisfying regulatory requirements, consideration could be given to: (i) monthly or even weekly sequential sampling after 30 minutes stagnation at selected homes and (ii) random daytime sampling, the latter providing an opportunity for acquiring a representative amount of information on pipe-work circumstances.

There is scope for further research into the behavioural characteristics of sequential sampling, involving a coupling of experimental work with model development.

Chapter 10

Conclusions

(1) Computational compliance modelling has been shown to have an important potential role in the optimisation of plumbosolvency control in both Canada and the United States. Modelling tools can also support investigations at individual homes, if the specific calibration data are available.

(2) CFD analysis and the modelling tools can support the field characterisation of the skewing and lowering of results from sequential sampling; such effects will need to be considered when sequential samples are used in the optimisation of plumbosolvency control.

(3) Establishing adequate calibration data on pipe-work characteristics should not be particularly onerous or costly, particularly if opportunities to undertake brief inspections of service connections and premise plumbing are taken during sampling campaigns.

(4) Premise plumbing characteristics are critical to the results of stagnation sampling.

(5) Plumbosolvency characteristics also strongly influence the results of sampling for lead in drinking water and can be quickly determined by laboratory testing.

(6) Computational compliance modelling has given a preliminary indication that there is scope to further reduce lead concentrations in City A by orthophosphate dosing, but operational feasibility would first need to be confirmed.

(7) Computational compliance modelling indicates that the current plumbosolvency control treatment measures in City B are close to optimum by pH elevation alone.

(8) A diagnostic assessment of water quality data, sequential survey results and the history of the installation of lead service lines, when taken together, suggest that lead in drinking water may be an issue in City C that warrants further prompt investigation.

(9) Sequential sampling results after both 30 minutes and 6 hours stagnation are strongly influenced by pipework characteristics.

(10) There is scope to improve the various sampling protocols that are used in the regulation of lead in drinking water in both Canada and the United States. Computational modelling tools can greatly assist the development of robust sampling protocols via exhaustive behavioural assessment.

Chapter 11

References

Cardew P. T. (2009). The impact of ortho-phosphate treatment on lead in drinking water in North West of England. Proceedings of International Conference, COST Action 637, October 2008, Lisbon.

Cartier C., Laroche L., Deshommes E., Nour S., Richard G., Edwards M. and Prevost M. (2011). Investigating dissolved lead at the tap using various sampling protocols. *Journal of the American Water Works Association*, **103**(3), 55–67.

Croft N., Pericleous K. and Cross M. (1995). PHYSICA: a multiphysics environment for complex flow processes. In: *Numerical Methods in Laminar and Turbulent Flows Vol IX*, C. Taylor *et al.* (eds), Pineridge Press, Swansea, pp. 1269–1280.

Drinking Water Inspectorate (1992). *Nitrate, Pesticides & Lead 1989 and 1990.* Department of the Environment, Welsh Office, London.

Drinking Water Inspectorate (1996). *Nitrate, Pesticides & Lead 1991 to 1994.* Department of the Environment, Welsh Office, London.

Drinking Water Inspectorate (2000). Determination of requirements to meet new lead standards. Information Letter 12/2000.

Drinking Water Inspectorate (2001). Further guidance on requirements to meet new lead standards. Information Letter 3/2001.

Drinking Water Inspectorate (2010). Guidance document. Guidance on the implementation of the Water Supply (Water Quality) Regulations 2000 (as amended) in England. September 2010.

Drinking Water Inspectorate (2011). Regional water quality reports for 2005 to 2009. Last consulted 1 October 2012. Available from www.dwi.gov.uk

Hayes C. R. (2002). Computational Modelling of Lead in Drinking Water. PhD thesis, University of Wales, Swansea.

Hayes C. R. (2008). Optimisation tools for achieving the lead standard of 10 µg/l in drinking water. Proceedings of an International Conference on Metals and Related Substances in Drinking Water, COST Action 637, October 2007, Antalya, Turkey.

Hayes C. R. (2009). Computational modelling to investigate the sampling of lead in drinking water. *Water Research*, **43**, 2647–2656.

Hayes C. R. and Hydes O. D. (2012). UK experience in the monitoring and control of lead in drinking water. *Journal of Water and Health*, **10**(3), 337–348.

Hayes C. R., Bates A. J., Jones L., Cuthill A. D., Van der Leer D. and Weatherill N. P. (2006). Optimisation of plumbosolvency control using a computational model. *Water and Environment Journal*, **20**, 256–264.

Hayes C. R., Incledion S. and Balch M. (2008). Experience in Wales (UK) of the optimisation of ortho-phosphate dosing for controlling lead in drinking water. *Journal of Water and Health*, **06.2**, 177–185.

Health Canada (1992). Guidelines for Canadian Drinking Water Quality.

Health Canada (2009). *Guidance on Controlling Corrosion in Drinking Water Distribution Systems.* Health Canada, Ottawa. ISBN 978-1-100-14193-0.

Hepple S. (2009). Metal Leaching from Pipes and Components in Contact with Drinking Water. MRes thesis, Swansea University.

International Water Association (2010). *Best Practice Guide on the Control of Lead in Drinking Water.* IWA Publishing, London. ISBN 9781843393696.

Kuch A. and Wagner I. (1983). A mass transfer model to describe lead concentrations in drinking water. *Water Research*, **17**(10), 1303–1307.

Ontario Ministry of the Environment (2007). Ontario Regulation 399/07, amended to 170/03.

Rhie C. M. and Chow W. L. (1983). Numerical study of the turbulent flow past an airfoil with trailing edge separation. *AIAA Journal*, **21**, 1525–1532.

Scottish Centre for Infection and Environmental Health (2000). Scottish New Homes Lead Survey, Stage 1. October 2000.

Van der Leer D., Weatherill N. P., Sharp R. J. and Hayes C. R. (2002). Modelling the diffusion of lead into drinking water. *Applied Mathematical Modelling*, **26**(6), 681–699.

van Dormal J. P. and Raithby G. D. (1984). Enhancements to the SIMPLE method for predicting incompressible fluid flows. *Numerical Heat Transfer*, **7**, 147–163.

Versteeg H. K. and Malalasekera W. (1995). *An Introduction to Computational Fluid Dynamics–The Finite Volume Method.* Prentice Hall, Harlow.

Appendix 1
Calibration data

CITY A

Table A1 Lead pipe length distribution.

Meters	%	Meters	%
1	0.74	21	0.39
2	4.03	22	0.25
3	5.80	23	0.26
4	8.76	24	0.10
5	10.80	25	0.11
6	10.10	26	0.05
7	8.66	27	0.04
8	9.25	28	0.05
9	7.87	29	0.02
10	6.84	30	0.04
11	5.84	31	0.02
12	4.72	32	0.01
13	3.82	33	0.01
14	3.17	34	0.01
15	2.37	35	0.01
16	2.07	41	0.01
17	1.35	43	0.01
18	0.93	52	0.01
19	0.93	55	0.01
20	0.53	58	0.01

Table A2 Diameter distribution.

Diam (mm)	%
12.5	0.37
15.8	90.58
18.8	6.38
25	1.77
31.3	0.85
37.5	0.05

Table A3 Non-lead pipe length distribution.

Meters	%
0.0	5
2.5	15
5.0	27
7.5	22
10.0	16
15.0	10
20.0	3
25.0	2

Table A4 Correction factors for premise plumbing in Providence.

Pb (µg/l)	%
2.5	72.1
6.3	8.8
8.8	4.3
11.3	3.6
13.8	3.4
16.3	2.6
18.8	2.4
25.0	1.3
35.0	0.9
45.0	0.4
55.0	0.2

CITY B

Table B1 Lead pipe length distribution.

Meters	%
2.5	3
5.0	7
7.5	10
10.0	40
12.5	20
20.0	10
24.0	7
30.0	2
40.0	1

Table B2 Lead pipe and non-lead pipe diameter distributions.

Lead service lines		Premise plumbing	
%	mm	%	mm
2	25	0	25
10	18	5	18
78	15.6	0	15.6
10	12	95	12

Table B3 Non-lead pipe length distribution.

Meters	%
0	3
6	7
9	9
12	11
15	20
18	20
21	12
24	8
27	5
30	3
35	1
38	1

Appendix 2

Examples of model output

```
# ============================================================================
# ||                                                                        ||
# ||          OVERVIEW OUTPUT FILE FOR US SAMPLING SIMULATION               ||
# ||                                                                        ||
# ||                   Produced by LCS Version 1.606b                       ||
# ||                                                                        ||
# ============================================================================
#
#     Name Of Zone                                  :
#           EXAMPLE OF ZONAL COMPLIANCE OUTPUT
#
#     Networked Zone                                :    False
#
#     Continuous Variables                          :    False
#
#     Variable Maximum Exchange Rate                :    False
#
#     Date Of Zone Generation                       :    22/04/2012 11:06:23
#
#     Date Of Zone Calculation                      :    22/04/2012 11:09:57
#
#     Number Of Supplies In Zone                    =        10000
#
#     Number Of Supplies With Lead                  =         5000
#
#     Maximum Exchange rate (ug/s m^2)              =     5.000E-02
#
#     Equilibrium Concentration (ug/l)              =      75.0
#
#     Flow rate (l/s)                               =     0.100
#
#     Number Of water pattern files                 =            5
#
#     Name for water usage pattern file 1           :    std-patt-1
#
#     Name for water usage pattern file 2           :    std-patt-2
#
#     Name for water usage pattern file 3           :    std-patt-3
#
#     Name for water usage pattern file 4           :    std-patt-4
#
#     Name for water usage pattern file 5           :    std-patt-5
#
#     Lead Pipe Distribution File                   :    pb-len-prov
#
#     Copper Pipe Distribution File                 :    cu-len-prov
#
#     Lead Pipe Diam Distribution File              :    diam-prov
#
#     Copper Pipe Diam Distribution File            :    diam-prov
#
#     Cu and Pb diams same at an address            =            1
#
#     Total Water Distribution File                 :    tot-vol
#
#     Water Pattern types surveyed                  :    A
#
#     Number Of Time Ranges                         =            1
#
#     Time Range 1                                  :     09-00 to 17-00
#
#     Premises lead option, 0 implies no premises=lead             1
#
#     Premises lead Dist. File                      :    premises
#
#     Scale factor for premises lead                =       1.00
#
#     Number Of Samples Taken                       =          100
#
#     Number Of Surveys Performed                   =           10
#
#        Statistics For    6 Hour Stagnation At (     15.00ug/l) Standard
#
#        Failures and 90th percentile conc. for each litre sample
#
#                     Litre  1            Litre  2            Litre  3            Litre  4
 Survey :   1      6       6.300     24      71.163     28      66.860     18      67.519
 Survey :   2     17      38.470     44      71.163     37      73.047     24      52.931
 Survey :   3     13      38.470     41      71.163     34      73.045     23      72.511
 Survey :   4     11      16.300     37      71.163     28      71.936     15      31.668
 Survey :   5      9      11.776     31      71.163     30      71.936     20      65.543
 Survey :   6     16      38.470     33      71.163     32      73.045     19      66.028
 Survey :   7     14      23.672     39      71.163     37      73.045     26      65.562
 Survey :   8     15      26.306     39      71.163     27      73.045     20      52.336
 Survey :   9     20      38.470     44      71.264     38      73.045     23      67.519
 Survey :  10     13      27.226     29      71.163     30      73.045     25      59.784

 Average   :     13.40    26.546   36.10    71.173   32.10    72.205   21.30    60.140

 % Failure :     13.40    80.000   36.10   100.000   32.10   100.000   21.30   100.000

 Minimum   :      6.00     6.300   24.00    71.163   27.00    66.860   15.00    31.668

 Maximum   :     20.00    38.470   44.00    71.264   38.00    73.047   26.00    72.511

 Std Dev   :      3.83    11.448    6.32     0.030    3.94     1.835    3.29    11.309
```

```
# ==========================================================================
# ||                                                                      ||
# ||          OVERVIEW OUTPUT FILE FOR SINGLE SUPPLY SIMULATION           ||
# ||                                                                      ||
# ||                  Produced by LCS Version 1.606b                      ||
# ||                                                                      ||
# ==========================================================================
```

Length of Lead Pipe (m) = 0.225E+02

Diameter of Lead Pipe (mm) = 0.180E+02

Length of Copper Pipe (m) = 0.157E+02

Diameter of Copper Pipe (mm) = 0.120E+02

Total Volume of Water Used During Simulation (l) = 0.450E+03

Flow Rate when tap is turned on (l/s) = 0.100E+00

Maximum exchange rate for Lead into water (ug/ m^2 .s)= 0.500E-01

Equilibrium value for Lead concentration (ug/l) = 0.750E+02

Lead concentration limit (ug/l) = 0.100E+02

Filname used for pattern of water usage throughout simulation
std-patt-2

Time Period	Flow	End Of Lead				End Of Copper				1L CFSTR			
From - Until	In S	Min Conc	Max Conc	Avg Conc	>Lim In S	Min Conc	Max Conc	Avg Conc	>Lim In S	Min Conc	Max Conc	Avg Conc	>Lim In S
00-00 hrs -00-29 hrs	0	0.0	0.0	0.0	0	0.0	0.0	0.0	0	0.0	0.0	0.0	0
00-30 hrs -00-59 hrs	0	0.0	0.0	0.0	0	0.0	0.0	0.0	0	0.0	0.0	0.0	0
01-00 hrs -01-29 hrs	0	0.0	0.0	0.0	0	0.0	0.0	0.0	0	0.0	0.0	0.0	0
01-30 hrs -01-59 hrs	0	0.0	0.0	0.0	0	0.0	0.0	0.0	0	0.0	0.0	0.0	0
02-00 hrs -02-29 hrs	0	0.0	0.0	0.0	0	0.0	0.0	0.0	0	0.0	0.0	0.0	0
02-30 hrs -02-59 hrs	0	0.0	0.0	0.0	0	0.0	0.0	0.0	0	0.0	0.0	0.0	0
03-00 hrs -03-29 hrs	0	0.0	0.0	0.0	0	0.0	0.0	0.0	0	0.0	0.0	0.0	0
03-30 hrs -03-59 hrs	0	0.0	0.0	0.0	0	0.0	0.0	0.0	0	0.0	0.0	0.0	0
04-00 hrs -04-29 hrs	0	0.0	0.0	0.0	0	0.0	0.0	0.0	0	0.0	0.0	0.0	0
04-30 hrs -04-59 hrs	0	0.0	0.0	0.0	0	0.0	0.0	0.0	0	0.0	0.0	0.0	0
05-00 hrs -05-29 hrs	0	0.0	0.0	0.0	0	0.0	0.0	0.0	0	0.0	0.0	0.0	0
05-30 hrs -05-59 hrs	0	0.0	0.0	0.0	0	0.0	0.0	0.0	0	0.0	0.0	0.0	0
06-00 hrs -06-29 hrs	113	0.6	72.6	37.1	58	0.0	72.6	37.0	59	0.0	72.4	36.8	75
06-30 hrs -06-59 hrs	113	0.6	17.1	9.0	57	0.6	17.1	9.0	57	0.9	17.0	9.1	55
07-00 hrs -07-29 hrs	225	0.6	16.1	4.6	57	0.6	16.1	4.6	57	0.6	16.1	4.6	53
07-30 hrs -07-59 hrs	225	0.6	16.1	4.6	57	0.6	16.1	4.6	57	0.6	16.1	4.6	53
08-00 hrs -08-29 hrs	338	0.6	15.1	3.1	57	0.6	15.1	3.1	56	0.6	15.1	3.1	52
08-30 hrs -08-59 hrs	338	0.6	15.1	3.1	57	0.6	15.1	3.1	56	0.6	15.1	3.1	52
09-00 hrs -09-29 hrs	225	0.6	16.1	4.6	57	0.6	16.1	4.6	57	0.6	16.1	4.6	53
09-30 hrs -09-59 hrs	225	0.6	16.1	4.6	57	0.6	16.1	4.6	57	0.6	16.1	4.6	53
10-00 hrs -10-29 hrs	113	0.6	17.1	9.0	57	0.6	17.1	9.0	57	0.6	17.0	8.9	55
10-30 hrs -10-59 hrs	113	0.6	17.1	9.0	57	0.6	17.1	9.0	57	0.7	17.0	9.0	55
11-00 hrs -11-29 hrs	68	0.6	17.5	14.8	57	0.6	17.5	13.1	50	0.7	17.4	10.7	42
11-30 hrs -11-59 hrs	68	0.6	17.5	14.8	57	0.6	17.5	14.8	57	6.5	17.4	14.8	59
12-00 hrs -12-29 hrs	113	0.6	17.1	9.0	57	0.6	17.5	10.0	64	1.0	17.4	11.5	71
12-30 hrs -12-59 hrs	113	0.6	17.1	9.0	57	0.6	17.1	9.0	57	0.7	17.0	9.0	55
13-00 hrs -13-29 hrs	225	0.6	16.1	4.6	57	0.6	16.1	4.6	57	0.6	16.1	4.6	53
13-30 hrs -13-59 hrs	225	0.6	16.1	4.6	57	0.6	16.1	4.6	57	0.6	16.1	4.6	53
14-00 hrs -14-29 hrs	113	0.6	17.1	9.0	57	0.6	17.1	9.0	57	0.6	17.0	8.9	55
14-30 hrs -14-59 hrs	113	0.6	17.1	9.0	57	0.6	17.1	9.0	57	0.7	17.0	9.0	55
15-00 hrs -15-29 hrs	45	17.7	17.7	17.7	45	0.6	17.7	10.9	27	0.7	16.6	7.4	19
15-30 hrs -15-59 hrs	45	17.7	30.8	21.2	45	17.7	30.8	21.2	45	16.7	26.7	20.6	45
16-00 hrs -16-29 hrs	45	17.7	30.8	21.2	45	17.7	30.8	21.2	45	18.0	26.8	21.2	45
16-30 hrs -16-59 hrs	45	17.7	30.8	21.2	45	17.7	30.8	21.2	45	18.0	26.8	21.2	45
17-00 hrs -17-29 hrs	68	0.6	30.6	17.2	57	17.5	30.6	19.9	68	17.7	26.7	20.2	68
17-30 hrs -17-59 hrs	68	0.6	17.5	14.8	57	0.6	17.5	14.8	57	6.6	17.6	14.8	59
18-00 hrs -18-29 hrs	225	0.6	16.1	4.6	57	0.6	17.5	5.1	64	0.6	17.4	5.8	70
18-30 hrs -18-59 hrs	225	0.6	16.1	4.6	57	0.6	16.1	4.6	57	0.6	16.1	4.6	53
19-00 hrs -19-29 hrs	113	0.6	17.1	9.0	57	0.6	17.1	9.0	57	0.6	17.0	8.9	55
19-30 hrs -19-59 hrs	113	0.6	17.1	9.0	57	0.6	17.1	9.0	57	0.7	17.0	9.0	55
20-00 hrs -20-29 hrs	90	0.6	17.3	11.2	57	0.6	17.3	11.2	57	0.7	17.2	10.9	55
20-30 hrs -20-59 hrs	90	0.6	17.3	11.2	57	0.6	17.3	11.2	57	1.2	17.2	11.2	55
21-00 hrs -21-29 hrs	68	0.6	17.5	14.8	57	0.6	17.5	13.1	50	1.2	17.4	11.1	42
21-30 hrs -21-59 hrs	68	0.6	17.5	14.8	57	0.6	17.5	14.8	57	6.5	17.4	14.8	59
22-00 hrs -22-29 hrs	45	17.7	17.7	17.7	45	0.6	17.7	13.6	34	6.5	17.4	13.7	36
22-30 hrs -22-59 hrs	45	17.7	30.8	21.2	45	17.7	30.8	21.2	45	17.0	26.7	20.6	45
23-00 hrs -23-29 hrs	23	17.8	30.9	24.8	23	17.7	30.9	20.7	23	18.0	22.6	19.0	23
23-30 hrs -23-59 hrs	23	31.1	31.1	31.1	23	17.8	31.1	24.8	23	21.2	26.9	24.1	23
Totals	4510			8.5	1913			8.4	1894			8.4	1851

Concentrations exiting lead pipe for 30 min stand test are:

Litre	1	2	3	4	5	6	7	8	9	10	Max
Minimum =	17.555	17.640	17.725	17.810	17.895	0.671	0.633	0.633	0.633	0.633	
Maximum =	17.632	17.717	17.802	17.886	17.971	18.030	0.636	0.633	0.633	0.633	18.030
Average =	17.594	17.679	17.764	17.848	17.933	13.223	0.634	0.633	0.633	0.633	17.933

Concentrations exiting copper pipe for 30 min stand test are:

Litre	1	2	3	4	5	6	7	8	9	10	Max
Minimum =	0.000	0.000	17.413	17.659	17.744	17.829	17.914	0.648	0.633	0.633	
Maximum =	0.000	16.777	17.651	17.736	17.821	17.905	17.990	18.024	0.636	0.633	18.024
Average =	0.000	3.960	17.593	17.698	17.783	17.867	17.952	9.329	0.634	0.633	17.952

Time for this flow = 10 seconds
Output was 1.000 litres.

IWA Publishing's authorised EU representative for General Product Safety Regulations is Diane D'Arras, 15 rue Duret, 75116 Paris, France, e-mail: safety@iwap.co.uk.

Printed and bound by CPI Group (UK) Ltd, Croydon, CR0 4YY
17/09/2026
02218798-0020